Xenobiotics and Food-Producing Animals

ACS SYMPOSIUM SERIES 503

Xenobiotics and Food-Producing Animals

Metabolism and Residues

D. H. Hutson, EDITOR
Shell Research Limited

D. R. Hawkins, EDITOR
Huntingdon Research Centre

G. D. Paulson, EDITOR
U.S. Department of Agriculture

C. B. Struble, EDITOR
Hazleton Laboratories

Developed from a symposium sponsored
by the Division of Agrochemicals of the American Chemical Society
and the International Society for the Study of Xenobiotics
at the Fourth Chemical Congress of North America
(202nd National Meeting of the American Chemical Society),
New York, New York,
August 25–30, 1991

American Chemical Society, Washington, DC 1992

Library of Congress Cataloging-in-Publication Data

Xenobiotics and food-producing animals: metabolism and residues / D. H. Hutson . . . [et al.], editor.

p. cm.—(ACS symposium series, ISSN 0097-6156; 503)

"Developed from a symposium sponsored by the Division of Agrochemicals of the American Chemical Society at the Fourth Chemical Congress of Noroth America (202nd National Meeting of the American Chemical Society), New York, New York, August 25–30, 1991."

Includes bibliographical references and indexes.

ISBN 0-8412-2472-2

1. Veterinary drugs—Congresses. 2. Xenobiotics—Metabolism—Congresses. 3. Veterinary drugs—Metabolism—Congresses. 4. Veterinary physiology—Congresses. 5. Veterinary drug residues—Congresses.

I. Hutson, D. H. (David Herd), 1935– . II. American Chemical Society. Division of Agrochemicals. III. American Chemical Society. Meeting (202nd: 1991: New York, N.Y.) IV. Series.

SF917.X46 1992
636.089'57—dc20 92-25718
CIP

The paper used in this publication meets the minimum requirements of American National Standard for Information Sciences—Permanence of Paper for Printed Library Materials, ANSI Z39.48-1984. ∞

PRINTED IN THE UNITED STATES OF AMERICA

Foreword

THE ACS SYMPOSIUM SERIES was first published in 1974 to provide a mechanism for publishing symposia quickly in book form. The purpose of this series is to publish comprehensive books developed from symposia, which are usually "snapshots in time" of the current research being done on a topic, plus some review material on the topic. For this reason, it is necessary that the papers be published as quickly as possible.

Before a symposium-based book is put under contract, the proposed table of contents is reviewed for appropriateness to the topic and for comprehensiveness of the collection. Some papers are excluded at this point, and others are added to round out the scope of the volume. In addition, a draft of each paper is peer-reviewed prior to final acceptance or rejection. This anonymous review process is supervised by the organizer(s) of the symposium, who become the editor(s) of the book. The authors then revise their papers according the the recommendations of both the reviewers and the editors, prepare camera-ready copy, and submit the final papers to the editors, who check that all necessary revisions have been made.

As a rule, only original research papers and original review papers are included in the volumes. Verbatim reproductions of previously published papers are not accepted.

M. Joan Comstock
Series Editor

Contents

INDEXES

Preface

GOOD ANIMAL HUSBANDRY IS CRITICAL in the conversion of plant-derived feedstuff into animal protein for human consumption. Bioactive chemicals play an important role in maintaining health, rapid growth rate, and efficient feed conversion in animals used for food production. Control of disease by therapeutic and prophylactic use of veterinary pharmaceutical products has gained in importance as the numbers of animals held in close proximity for confined rearing have steadily increased, thus magnifying risk of animal-to-animal disease transmission. However, improvements in the use of pharmaceutical products to control pests in low-intensity animal production are also important. For example, myasis-producing parasitic arthropods, such as the sheep blow fly *(Lucila cuprina)* and disease-spreading blood-sucking flies, when uncontrolled have devastating effects on the well-being of animals and on the economics of food production.

Thus, an array of bioactive compounds, often derived directly from the pharmaceutical and crop protection industries, are used: antibacterials, anticoccidials, miticides, nematicides, parasiticides (for control of both internal and external parasites), and insecticides. In addition, other specialty chemicals, such as growth-regulating, estrous-synchronizing, and nutrient-repartitioning agents, are in use and are being considered for use in animal production.

In the United States, crop protection agents used in food production are regulated by the Environmental Protection Agency, and the use of veterinary products is closely regulated by the Food and Drug Administration. Clear differences exist between the problems associated with the administration of veterinary products and the exposure of animals to crop protection agents. The use of veterinary products involves their deliberate and controlled application for benefit. Studies conducted in the development phase of a veterinary product address, in addition to the toxicology, the fate of the chemical in the test species. Withdrawal periods are set so that residues in edible tissues are at or below an acceptable concentration.

In contrast, the exposure of animals to crop protection agents is adventitious and usually via feedstuff. Animal studies on crop protection agents are designed to determine if residue transfer (from feed to animal product) occurs at an acceptably low level. However, there is much in common between studies on veterinary products and crop protection agents. Some of the agents (e.g., insecticides) are common. The species exposed to these agents are also common (ruminants, swine, poultry, and fish). Of paramount importance, the terminal residues to which humans are exposed via food consumption present similar toxicological issues.

Most of this book concerns the metabolism of veterinary products, as opposed to crop protection agents, because the former grouping covers a wider range of bioactivities. The content will, however, be of interest to those working with both types of agents. Specifically, a substantial amount of information about general methods for studying the fate of xenobiotics in species grown for human consumption is available. Increased attention is being paid to the fate of veterinary drugs in the environment, a major concern for crop protection agents for 30 years. Turning this situation around, environmental scientists concerned with the fate of other (e.g., industrial) chemicals should gain understanding of the capacity of food-producing animals to metabolize and eliminate chemicals. This knowledge is important as a basis for judgments about potential risk in the event of a major chemical contamination.

We acknowledge the support of the International Society for the Study of Xenobiotics and the American Chemical Society's Division of Agrochemicals. We thank the authors for their efforts and we are grateful for the help provided by the ACS Books Department.

D. H. HUTSON
Shell Research Limited
Kent, England

G. D. PAULSON
U.S. Department of Agriculture
Fargo, ND 58105–5674

D. R. HAWKINS
Huntingdon Research Centre
Cambridgeshire, England

C. B. STRUBLE
Hazleton Laboratories
Madison, WI 53707

April 6, 1992

Veterinary Drugs: Use and Regulation

Chapter 1

Uses and Regulation of Veterinary Drugs

Introduction

K. N. Woodward

Veterinary Medicines Directorate, Woodham Lane, New Haw, Addlestone, Surrey KT15 3NB, United Kingdom

There are a variety of veterinary drugs available and at the disposal of the veterinarian and the safety of these to the consumer must be assured in most countries of the world before marketing authorizations can be granted. A large body of pharmacological, toxicological and residues data is generated and assessed so that the toxicological profile of the drug can be established and a maximum residue limit elaborated. This, along with the residues depletion profile, allows a withdrawal period to be defined so that the consumer is protected from exceeding the acceptable daily intake for the drug in question. Worker safety is also of paramount importance when assessing a drug and its formulations prior to marketing authorisation. Veterinary medicines take many forms and a few of the more important groups are described in this article.

The range of veterinary medicines now available reflects both the diseases they are intended to combat and the range of species they are intended to treat. In addition, there are a number of drugs available for so-called zootechnical treatment (eg the use of steroid hormones in synchronization of oestrus) as opposed to disease treatment or prophylaxis. These aspects will be briefly discussed in this article, but a certain amount of liberty has been taken with the title so that the word "use" is interpreted in its widest sense to include a description of what occurs before "use" is allowed! Consequently, a large part of this work will deal with the general requirements of marketing authorization; the regulatory requirements that are applied before a veterinary medicine may be marketed.

Uses of Veterinary Medicines

Veterinary medicines, like their human counterparts come in many forms (1). Some would be instantly recognizable as tablets, pills

0097–6156/92/0503–0002$06.00/0

and injectables but others, because of clinical need, the species involved or the needs of herd or flock scale treatment, differ markedly from anything given to humans. The major groups of medicines however are roughly the same and these will be briefly described below, with the emphasis being on those used in food producing animals.

Antimicrobial and antibiotic agents. A large range of synthetic, semisynthetic and natural antibiotics are used for the treatment of infectious diseases in food producing animals. In the mid-1950s penicillin was seen as a general tool of salvation in both veterinary and human medicine, but now an array of ß-lactam antibiotics is available for a large range of indications. The other major categories are the tetracyclines, the aminoglycosides, the macrolides and polymixins (2-5).

Sulfonamides are the major class of antimicrobial agent (as opposed to antibiotics) and these can be viewed as derivatives of sulfanilamide, the archetypal sulfonamide. A range of compounds has been synthesised for fast, medium or long acting abilities (6-7). They are often formulated with folate antagonists such as trimethoprim to obtain what is claimed to be a synergistic action. Sulfamethazine (sulphadimidine) is perhaps the most widely used sulfonamide in veterinary medicine, especially in pig production where it is used to prevent repiratory disease (6). The use of this drug in pig production has given rise to problems of residues, particularly in the kidney, in both Europe and the USA. The reasons for this remain unclear for although failure to observe withdrawal periods plays a part, other contributing factors include ingestion of faeces from treated animals and contamination of untreated feed with medicated feed (8).

Ectoparasiticides. Cattle and sheep are particularly vulnerable to ectoparasites. In Central and South America, the USA and Europe, cattle are attacked by cattle grubs. In the Northern Hemisphere these "grubs" are the larval stages of the warble fly (*Hypoderma* spp) while in South America, larvae of tropical warbles (*Dermatobia hominis*) are the culprits (6). They are generally treated with pour-on formulations which tend to be viscous preparations often containing organophosphorus compounds. Sheep scab is a notifiable disease in both the USA and the United Kingdom, which results in the loss of the fleece. Although it is not generally regarded as a lethal disease, it poses severe animal welfare and economic problems. It is almost universally treated by dipping in aqueous solutions containing organophosphorus insecticides or synthetic pyrethroids (5).

Anthelmintics. There is insufficient scope in a paper of this kind to discuss the more interesting points of veterinary parasitology. Suffice it to say that food producing animals of all species are literally plagued by a large range of internal parasites resulting in distressing diseases and substantial economic losses (6,7). The interested reader is referred to the chapters by Roberson in the excellent work edited by Booth and

McDonald (31) which will serve as useful introductions. A variety of anti-parasitic drugs have been developed and three examples, the benzimidazoles, levamisole and ivermectin will be mentioned briefly here.

Thiabendazole was the first benzimidazole to achieve wide use. It is indicated for a range of infestations including those caused by *Haemonchus*, *Trichostrongylus* and *Strongyloides* species in cattle and sheep. Newer compounds include albendazole, oxfendazole, fenbendazole and mebendazole (9,10). In addition pro-drugs, exemplified by febantel, which undergo cyclisation *in vivo* to yield benzimidazoles have now been developed. Most of these drugs have some degree of teratogenic potential leading to concern about their residues, but a more practical concern has arisen over effects on the developing fetus in treated animals. Albendazole, cambendazole and parbendazole are teratogenic in sheep leading to specific contra-indications in pregnant animals, while other benzimidazoles are inactive in this respect.

Levamisole is highly effective against gastrointestinal nematodes and is widely used in cattle, sheep and pigs in addition to numerous other species. Levamisole is the l-isomer of dl-tetramisole. It appears to be the active isomer of the racemic mixture which is itself marketed as an antinematodal agent. Concerns have been expressed by Joint FAO/WHO Expert Committee on Food Additives (JECFA) over its apparent ability to induce agranulocytosis and neutropenia in humans given the drug for therapeutic purposes. This concern led the Committee to set a temporary acceptable daily intake (ADI) of 0-0.003 mg/kg body weight pending the results of further research on this phenomenon and its relevance to the safety assessment of levamisole residues (11).

Ivermectin is a macrolide compound derived from abamectin, a metabolite produced by *Streptomyces avermitilis*. More precisely, it is a mixture of two compounds, 22,23-dihydroavermectin B_{1a} (80%) and 22,23-dihydroavermectin B_{1b} (20%) (12). It is widely and successfully used for onchocerciasis treatment in human medicine and has found widespread use as a nematocidal and cestodocdal agent in veterinary medicine (12). Although it is claimed that it inhibits motility of the parasite by acting on γ-aminobutyric acid (GABA) receptors and by blocking chloride ions, the full mechanism is as yet not fully understood.

Antifungal agents. Several drugs are available as topical antifungal agents including thiabendazole, ketoconazole and a number of aliphatic acids such as undecylenic acid (4,13). Perhaps the two best known systemic drugs are nystatin and griseofulvin. Nystatin is remarkably low in toxicity when given orally but is much more toxic after parenteral administration. As it is poorly absorbed after oral administration it is useful for gastrointestinal tract infections. Griseofulvin, after absorption from the gastrointestinal tract is deposited in skin, hair and nails and is useful in the treatment of dermatomycoses. It is, however, teratogenic at high doses, at least in the cat, and its use in pregnancy is therefore contra-indicated in this and other species.

Steroid Hormones. Anabolic hormones like testosterone and its synthetic analogues such as trenbolone have been widely used in beef production for several years. The non-steroidal anabolic agent zeranol has also been widely used for this purpose (14). Growth promoting uses of all steroid hormones were recently banned by the European Community, but various zootechnical (eg synchronization of estrus) and therapeutic (eg prevention of abortion) uses of the endogenous hormones and their synthetic esters are permitted (15,16).

Somatotropins. Somatotropins are naturally occuring polypeptides found in all species although those intended for use in food production, largely to increase milk yields, may be synthesized using recombinant technology (17-19). There has been much controversy over the use of these materials ranging from the question of economic need to food safety concerns (20-22). The European Community's Committee for Veterinary Medicinal Products considered that at least one of these products did not present a risk to human health, and gave a positive opinion on the agent (23). It is impractical in an article of this type to list all the arguments for and against these drugs but the controversies involved seem set to rumble on for some time to come! Currently, the somatotropins are not authorised as milk yield enhancers in the USA or in the United Kingdom.

Fish farming. Although not a "therapeutic use", fish farming or aquaculture as it is now often called, is worthy of mention as it represents a new area of animal production and it poses its own problems. One of the major growth areas in fish farming is salmon culture. This is particularly suited to areas of the world where large inland expanses of both salt and fresh water are available, in relatively sheltered environments. These conditions allow the salmon farmer to grow the fish in their natural marine and fresh water environments thus reflecting their natural habitats.

Salmon farming is associated with two major diseases. Furunculosis, a bacterial disease caused by *Aeromonas salmonicida*, and sea-lice, an arthropod and an external parasite which attacks the surface of the fish (24,25). Both diseases are rapidly fatal and cause serious economic losses. In the United Kingdom, the only licensed medicine used to control sea-lice is a formulation containing the organophosphorus compound dichlorvos. Before this could be authorised, an enormous amount of ecotoxicology data had to be reviewed allowing for a risk assessment of dangers to the immediate aquatic environment, including the hazards to other arthropods. Furunculosis is prevented with the use of various antibiotics including amoxycillin and oxolinic acid (26). Again, environmental assessments were necessary before these drugs could be authorized. Although it is not the topic under discussion here, it is worth noting that there is also considerable environmental concern over the "natural" effluents from hatcheries (27).

Marketing Authorization

It would be impracticable to try to describe the various schemes established by regulatory agencies throughout the world to authorize users of veterinary medicines. Instead, the general requirements will be discussed. Usually, three main areas, pharmaceutical quality, efficacy and drug safety are examined and these must be satisfactory before marketing authorization is granted. The corner-stone of these three areas is product safety. A drug which is not of the correct quality because it contains toxic contaminants or because it is not sterile is not safe. Similarly, a drug which does not perform in the way described in the product literature is also not safe. The term "safety" speaks for itself: the product itself must be "safe" for the intended animal, it must not be hazardous to the person using it, it must not harm the environment and it must not leave potentially harmful residues in food intended for human consumption. In the public mind and in the corporate regulatory mind, it is drug safety which is of paramount importance when food producing animals are being considered. Food safety issues are currently the subject of intense debate in many areas of the world and many see pesticide and veterinary drug residues as posing potential threats to public health.

The safety of a veterinary drug to humans depends on a number of factors, some intrinsic to the animal being treated, some to the properties of the drug itself, and some to the method of use - or abuse of the veterinary medicine. These can conveniently considered under two main headings:

- safety to the consumer
- safety to the operator

Safety to the Consumer

For any chemical agent to exert a toxic effect in humans or in animals there are two important considerations - the toxic properties of the substance and the dose received. For veterinary drug residues, an assessment of safety involves an investigation of the toxicity of the drug and a quantitative study of the residues present in animal tissues.

Over the last two decades, a broadly accepted package of toxicity tests has emerged for assessing the toxicity of chemicals whatever their intended purpose. The tests are conducted in laboratory animals, usually rats and mice, and their objective is the identification of a dose level at which a toxic effect does not occur, the no-effect or no-observed effect level (NEL or NOEL) (28,29). From the results of these tests also help to establish **the toxicological profile of the chemical can also be established** - is it a teratogen, a genotoxic carcinogen or an uncoupler of oxidative phosphorylation for instance? Different regulatory authorities have differing requirements therefore it is impossible to compose a precise list of studies required for drug approval. Sometimes the requirement to conduct one test depends on the results of another. However, most authorities demand more or less the following:

- a study of acute toxicity in rodents
- a study of short-term toxicity - 28 or 90 days
- a battery of tests for genotoxicity
- an investigation of carcinogenic activity
- an investigation of teratogenic activity
- studies of reproductive performance.

Current scientific dogma claims that there is no safe level for a genotoxic carcinogen, an agent which causes cancer by a direct effect on the genetic material of a cell. Theoretically a single molecule could give rise to a mutation resulting in a cancer cell and then a cancer (30). Whatever the merits of this argument, it is widely regarded as unacceptable to be faced with the possibility of residues of a genotoxic carcinogen in food of animal (or any other) origin. Similar sentiments would apply for similar reasons, to genotoxic materials with the ability to affect the germ-line cells. Of course, if a genotoxic carcinogen is metabolized in the target animal to non-active residues, then an alternative risk assessment is possible and the drug will be viewed as more acceptable. This was in fact part of the evaluation of the drug carbadox, a growth promoter for pigs, by JECFA. It noted that the drug itself was both mutagenic and carcinogenic in laboratory studies but its residues were inactive and hence acceptable (11).

Assuming that there are no adverse manifestations such as genotoxic carcinogenicity, and assuming again that some toxic effects have been noted, an NEL should be identifiable providing that a suitable range of doses has been chosen for the toxicology studies. Once an NEL has been identified an acceptable daily intake (ADI) can be calculated using a suitable safety factor (2,31-33). There is considerable debate over the magnitude of this safety factor but the one usually chosen is 100. If the drug produces no toxic effects in laboratory species but some minor adverse reaction has been noted in humans, for example during use as a human medicine, a smaller factor, usually 10, may be employed. If the range of tests was limited and the results of dubious significance, or if the studies were poorly performed, a larger safety factor may be applied and a temporary ADI adopted. The ADI therefore as NEL/100 is usually quoted in terms of mg drug/kg body weight/day or mg/kg per day (32-36).

An ADI can also be calculated for a non-genotoxic carcinogen (one which operates via an epigenetic mechanism) providing that the mechanism of carcinogenicity is known. A good example is provided by the JECFA assessment of the sulfonamide drug sulfamethazine (sulphadimidine). This was shown to be a thyroid carcinogen in rodents but was accepted to be a non-genotoxic compound. Moreover, it was concluded that the mechanism of carcinogenicity involved perturbations of the thyroid-pituitary-hypothalamus axis, changes in thyroid hormone levels and a resulting hyperplasia of the thyroid. Although the full mechanism is still poorly understood, it was accepted that an NEL could be based on the thyroid effects and an ADI calculated (37). Similar approaches have been taken with steroid hormones where no-hormonal effect levels can be determined in suitable experimental models (38,39).

Having determined an ADI, it is essential that consumption of food of animal origin by humans will not result in this value being exceeded. This upper limit is known as the maximum residue limit or MRL (28,31). Its elaboration depends on a number of factors including the likely degree of consumption of the food commodity or commodities in question and the general quantities or disposition of residues in each tissue. JECFA has proposed daily intakes of food commodities; 300g muscle, 100g liver, 50g kidney, 50g fat and 1.5l of milk, which it uses to elaborate the MRL and ensure that the ADI is unlikely to be exceeded (11,35). There is currently much debate over whether these values represent a realistic food intake for the commodities involved. For example, do they take into account the so-called extreme consumer who might as an instance eat large daily quantities of liver? Do they represent international food consumption? Or, to put the question more plainly, does 300g muscle cover the USA consumption of beef and the third world consumption of beef? The answer is quite evidently "no". Nevertheless JECFA, through the Codex Alimentarius system attempts to recommend MRL values which will be universally applicable and for these reasons, some might say limitations, what might be seen as average values for food consumption must be adopted if a practical solution is to be found.

Having arrived at an MRL or MRLs for a commodity or commodities, it is next essential to ensure that the tissues of animals treated with veterinary medicines do not exceed these values. In practice this is not as simple as it might at first appear. Even for a single active ingredient, the requirements of therapy (and marketing advantages) dictate that numerous formulations administered by various routes must be developed and made available for the veterinarian and farmer. Consequently, residues depletion in the living animal will not be constant but will vary according to the formulation given and the route of administration. It goes almost without saying that the species too, or more correctly the metabolism in that species, will determine the rate of residues depletion for any given formulation administered by a specific route. This means that residues studies are generally required by regulatory authorities for each species using each formulation and route of administration. Thus, residues depletion below the MRL for each situation is ensured (28,29). Drug residues studies usually involve treating the animal in question with the drug using the intended route of administration, the intended formulations, usually at the highest recommended dose and the maximum duration of administration. Animals are then serially slaughtered so that residues depletion can be studied and the time taken to achieve levels below the MRL established (31).

These studies should, if conducted properly, show the residues depletion profile of the formulation under study and will reveal for example any re-emergence of residues because of entero-hepatic recirculation. The time taken for the residues to be depleted to below the MRL for each of the tissues of interest is then usually chosen as the withdrawal period or withholding period (or times) for that formulation. Usually the studies must be conducted in each of the indicated species although simpler and cheaper bioequivalence studies where pharmacokinetic profiles are examined and compared, may be used to evaluate the withdrawal period in other food-producing species.

Withdrawal periods can be a cause for dispute between companies and regulatory authorities, often for competitive marketing reasons. If two similar products are available for a particular therapeutic purpose, the veterinarian or farmer will usually choose the one with the shorter withdrawal periods so that if necessary the animal can be sent to slaughter at the earliest possible time after recovery. These considerations are extremely important for milk because it cannot be kept until residues have depleted to below the MRL and milking cannot be postponed. Contaminated milk has to be discarded, thus attracting financial penalties these being all the more important the longer the withdrawal period. Drugs with shorter withdrawal periods offer an obvious advantage. Similar considerations can be applied to honey. When bees are treated for disease conditions, the drug accumulates in the honey (40-42). Here, it may slowly change to non-biologically active residues (43-45) but it may persist and the honey will need to be discarded until treatment has finished.

Treating fish poses different technical and therapeutic problems some of which will be mentioned in a later part of this Chapter. From a residues point of view, a particular problem arises from the general physiological processes peculiar to poikilotherms (46). Their metabolic rates are partly governed by their body temperature which is dependent upon the ambient temperature of the water in which they live; the cooler the water, the longer residues depletion takes. For this reason, residues depletion studies in fish are usually conducted at several temperatures chosen to represent the range of temperatures to which they will be exposed under farming conditions. Withdrawal periods are then quoted in degree days these being a function of both time and temperature.

Specific problems are raised by bound residues. After an animal has been treated with a medicine, its residues are present in plasma and tissues as parent drug and metabolite or metabolites. The residues may be present as free drug and/or metabolites or as covalently bound residues. This then raises the question of the degree of bioavailability of these bound residues and their biological activity (47,48). In most cases, this question has no simple answers. Of course a drug may be metabolized to carbon dioxide and hence bicarbonate ion or some other simple precursor of normal endogenous biochemicals. If these arise from a radiolabelled portion of the molecule, measurements of residues simply as incorporated radiolabel will lead the investigator to suspect bound residues when in fact there are only normal bodily constituents containing incorporated isotope.

If bound residues are found, their impact on the ADI must be assessed; can they be ignored or are they of toxicological significance? JECFA has addressed this issue and has recommended a systematic approach to the problem (11,37). JECFA suggests use of a mild extraction procedure to determine those residues which are clearly bioavailable. This is followed by a more vigorous extraction using acids or enzymic techniques to assess whether potentially biologically active compounds may be released in vivo. These studies can be backed-up by relay methodologies (residues transfer studies) whereby tissues from treated animals are fed to laboratory species and the release of drug-related moities measured, e.g. in plasma (47,49,50). The Committee stressed the need to treat each drug on a case-by-case basis (rather than laying down strict protocols to deal with bound residues as a common issue). The complexity of the problem can easily be seen by reference to the work of Lu and colleagues in the United States (51). They investigated bound residues arising from use of ronidazole, a nitroimidazole drug structurally related to metronidazole. Among other discoveries, they found that ronidazole covalently bound to proteins in the pig forming an adduct which they investigated after acid hydrolysis. Ronidazole itself is mutagenic but Lu and his co-workers demonstrated that the bound residues were devoid of genotoxic potential and so did not offer a risk to the consumer. A similar analysis has been made for residues of furazolidone (52).

The current FDA Guidelines outline a series of short-term and in vitro tests for the safety assessment of bound residues together with their chemical characterisation (53). A study of reversibility of adduct formation is also included and as with the JECFA recommendation, drugs are investigated in an individual manner. It seems likely that the JECFA and FDA approaches, at least in general terms, will become widely adopted in this particular area of hazard and risk assessment.

Microbiological Risk Assessment

Microbiological risk in this context refers to the possible effects of residues of antimicrobial drugs on the gut flora in humans exerting a selective pressure favoring either the growth of microorganisms with natural resistance to the drug in question, or the growth of microorganisms with acquired resistance. This is a very controversial area of risk assessment, for while the phenomenon of induction of resistance is well known (54-58), there is no evidence for the supposed effects in humans in vivo as a result of ingestion of food containing veterinary drug residues (59-61). While it is widely accepted that it would be virtually impossible to detect a toxic effect in humans due to drug residues because of the wide background of disease and the difficulties involved in attributing any effect to residues, emerging drug resistance should be more evident. Many antimicrobial and antibiotic substances are of low toxicity and MRLs therefore are often quite generous. However, applying the results of testing for antimicrobial resistance may result in very low MRLs and long withdrawal periods.

There are three basic types of study available (11,62,63):

- studies in human volunteers
- studies in germ-free (holoxenic) rodents
- *in vitro* studies with bacterial populations

The first of these involves the examination of the human fecal flora before and after treatment with antibiotics. Colonisation of the gastrointestinal tract, e.g. the oral cavity, by adventitious microorganisms is also investigated. The use of germ-free rodents models the human situation. These animals are inoculated with human gut flora and the effects of antibiotics and antimicrobials can then be studied. *In vitro* investigations examine the effects of varying concentrations of the drug or drugs of interest on cultures of indicator organisms. All these studies can be used to derive no-effect levels for toxicity towards the bacteria employed. More specifically, the minimum inhibitory concentration (MIC) values can be determined.

An example of this type of assessment is provided by the JECFA deliberations on oxytetracyline at its meeting in Rome in 1990 (11). The Committee noted that the toxicological potential of oxytetracyline was low but studies were available on its antimicrobial effects in dogs and human volunteers and a no-effect dose of 2mg per day was identified from the volunteer experiments. This led to an ADI of 0-0.003mg/kg body weight using a safety factor of 10. The toxicological studies would have given an ADI of around 0.18 mg/kg body weight using a safety factor of 100.

The induction of antibiotic resistance has been recognised for many years, and there is some evidence to suggest that antibiotic resistance may develop in pathogens in animals given antibiotics (58,60). These resistant pathogens may then be transmitted to humans. There is however no evidence at the moment that antimicrobial residues in meat or other animal products may lead to effects on the human gut flora and more research is obviously required in this area before major regulatory decisions are made. In the meantime, many will see the calculation of ADIs based upon microbiological data as an interim and added safety measure, whilst others will regard it as dubious science (59).

Operator Safety

Although this aspect is irrelevant to the assessment of residues and their implications for consumer safety, no account of the use of veterinary drugs would be complete without some passing mention which will restrict itself to considerations of safety in use of medicines rather than safety in their manufacture. Many formulations used in veterinary medicine offer very little scope for significant occupational exposure. Tablets and capsules for example ensure that occupational exposure is minimal, if not non-existent. However, operator exposure does occur and means need to be taken to reduce this. Organophosphorus compounds are used in veterinary medicines as ectoparasiticides in food-producing and companion animals; in several countries, including the UK, these have replaced organochlorine compounds. Companion animal products are exemplified by slow release collars used for the control of fleas and other parasites in cats and dogs. Here, it is important to ensure that the product does not allow the rapid release of large quantities of the active ingredient which would otherwise pose a serious risk to the owners. Sheep-dips and warblicide formulations contain such ingredients as chlorfenvinphos, diazinon and propetamphos. Apart from the obvious toxic effects of anti-cholinesterases (64) concern has been expressed over possible long-term effects following occupational exposure to organophosphorus compounds (65-71). In the United Kingdom, there is a comprehensive adverse reactions reporting system which covers suspected adverse reactions in both the animal patient and in humans using the medicines (72). This has revealed a small series of suspected adverse reactions to sheep dips which some have attributed to the organophosphorus component. These suspected adverse reactions have included wheezing, coughs, influenza-like effects and headaches (72). At the present time it is unclear if these are due to the active ingredients, to other excipients such as organic solvent or other, unknown factors. In the United Kingdom and other European Community Member States, a review of veterinary medicines is currently taking place under European Community legislation. This requires that many products currently on the market are assessed for safety, quality and efficacy as if they were new marketing authorisation applications (73) and as part of this review, the products containing organophosphorus compounds will be rigorously scrutinized to determine whether new precautions or adjustments to the formulations can be made to reduce the frequency of adverse reactions.

Another example of occupational problems associated with veterinary medicines is that of self-injection of oil-based vaccines (72). This can give rise to vascular compression, ischaemia and tissue damage, particularly if injection into the tendon sheath occurs (74). In the United Kingdom, this has resulted in advice being provided to hospital emergency departments so that prompt and appropriate surgical treatment can be given (75).

Many antibiotic formulations are given in feed and as many antibiotics have allergenic properties, problems can arise in their use. Regulatory authorities now insist that such formulations are rendered dust-free by admixture with inert oils or by the production of granular formulations so that occupational exposure is minimized (76).

In general, sensible occupational precautions, can be taken so that occupational exposure to veterinary medicines is reduced or excluded. The use of impervious gloves for spreading pour-on formulations, overalls for dealing with sheep-dips and face-masks or respirators for handling dusty products can all reduce human exposure. It is also essential for basic occupational hygiene advice to be given on the product or in the product literature so that informed users can take the necessary precautions or in extreme situations, use a suitable alternative (76).

Summary

It would not be effective in any way to review all the different types of medicines available to the veterinary surgeon, nor indeed the various dosage forms available and the methods of administration. The reader is referred to other works for that information. Suffice it to say that a variety of active ingredients and increasingly sophisticated formulations are now supplied by pharmaceutical companies. The onus is on their toxicologists, pharmacologists and residues experts to ensure that their labours guarantee so far as is reasonably practicable, that these products are safe for the consumer and for those otherwise exposed to them, including the veterinarian and farmer.

Literature cited

1. Blodinger, J. In Formulation of Veterinary Dosage Forms; Blodinger, J., Ed.; Marcel Dekker: New York, 1983; pp 135-173.
2. Huber, W.G. In Veterinary Pharmacology and Therapeutics; Booth, N.H.; McDonald, L.E., Eds; Iowa State University Press:Ames, Iowa, 1988; pp 813-821.
3. Huber, W.G. In Veterinary Pharmacology and Therapeutics, Booth, N.H.; McDonald LE. Eds.; Iowa State University Press:Ames, Iowa, 1988; 822-848.
4. Brander, G.C. In Chemicals for Animal Health Control; Taylor and Francis : London, 1986 and references therein.
5. Brander, G.C.; Pugh, D.M.; Bywater, R.J.; In Veterinary Applied Pharmacology and Therapeutics; Balliere Tindall: London, 1982; pp 356-433.
6. Anon. In The Merck Veterinary Manual; 6th Ed.; Merck and Co. Inc.: Rahway, 1986.
7. Bevill, R.F. In Veterinary Pharmacology and Therapeutics; Booth N.H.; McDonald L.E. Eds.; Iowa State University Press:Ames; Iowa, 1988; pp 785-795.
8. McCaughey, W.J.; Elliott C.T.; Crooks S.R.H. Vet. Rec., 1990, 126, 351-354.
9. Roberson, E.L. In Veterinary Pharmaology and Therapeutics; Booth, N.H.; McDonald, L.E. Eds.; Iowa State University Press:Ames; Iowa, 1988 pp 877-881.
10. Roberson, E.L. In Veterinary Pharmacology and Therapeutics; Booth, N.H.; McDonald, L.E. Eds.; Iowa State University Press:Ames: Iowa; 1988 pp 928-949.
11 Joint FAO/WHO Expert-Committee on Food Additives. Evaluation of Certain Veterinary Drug Residues in Food. Technical Report Series 799, WHO:Geneva; 1990.
12. Fink, D.W.; Porras A.G. In Ivermectin and Abamectin; Campbell, W.C.; Ed.; Springer-Verlag, London, 1989; pp 113-130 and references therein.

13. Huber, W.G. In *Veterinary Pharmacology and Therapeutics*; Booth, N.H.; McDonald, L.E. Eds.; Iowa State University Press:Ames, Iowa, 1988; pp 849-860.
14. Lindsay, D.G. *Food Chem. Toxicol*, 1985, *23*, 767-774.
15. Editorial. *Vet. Rec.*, 1986, *118*, 522-524.
16. Anon - Council Directive of 7 March 1988; 88/146/EEC. Offic J. Europ Commun., 1988, No L70, 16-18.
17. Vernon R.G.; Flint D.J. In *Biotechnology in Growth Regulaton*; Heap, R.B.; Prosser, C.G.; Lamming, G.E., Eds.; Butterworths: London, pp 57-71 and references therein.
18. Chilliard, Y. In *Use of Somatropin in Livestock Production*; Sejrsen, K., Vestergaard, M.; Neimann-Sorensen, A., Eds; Elsevier Applied Science: London, pp 61-87 and references therein.
19. Burrenich, C.; Vandeputte-van Messom, G.; Roets, E; Fabry, J.; Massart-Leen, A.-M. In *Use of Somatotropin in Livestock Production* Sejrsen, K.; Vestergaard, M.; Neimann-Sorensen, A. Eds.; Elsevier Applied Science: London, pp 377-280.
20. Hardin P. *The Milkweed*, February 1990.
21. Epstein S.S. *Ecologist*, 1989, *19*, 191-195.
22. Juskevich J.C.; Guyer C.G. *Science*, 1990, *249*, 875-884
23. Anon. *Animal Pharm* 12th April 1991, 1.
24. Roberts, R.J.; Shepherd, C.J. In *Handbook of Trout and Salmon Diseases*, Fishing News Books: Oxford; 1986, pp. 46-87.
25. Stuart, N. *In Practice*, March 1988, 47-53.
26. Liao, P.B. *Water and Sewage Works*, 1970, *117*, 291-297.
27. National Office of Animal Health (NOAH). In *Compendium of Data Sheets for Veterinary Products 1990-91*, Datapharm Publications Limited: London, 1990.
28 Woodward, K. N. In *Food Contaminants: Sources and Surveillance*; Creaser, C.; Purchase, R., Eds.; Royal Society of Chemistry: London, 1991, pp 99-108, and references therein.
29. Woodward, K. N. *Biologist*, 1991, *38*, 105-108, and references therein.
30. Somogyi, A. In *Carcinogenic Risks. Strategies for Intervention*, IARC Scientific Publications, No. 25, IARC: Lyon, 1979; pp 123-127.
31. Booth, N. H. In *Veterinary Pharmacology and Therapeutics*; Booth, N.H.; McDonald, L. E. Eds.; Iowa State University Press: Ames; Iowa, 1988; pp 1149-1205.
32. International Programme on Chemical Safety. In *Principles for the Safety Assessment of Food Additives and Contaminants in Food*, Environmental Health Criteria 70, WHO: Geneva, 1987, pp.75-86.
33. International Programme on Chemical Safety. In *Principles for the Toxicological Assessment of Pesticide Residues in Food*, Environmental Health Criteria 104, WHO: Geneva, 1990, pp.76-83.
34. Bigwood, E. J. *CRC Crit.Rev. Toxicol.*, 1973, *2*, 41-93.
35. Perez, M. K. *J. Toxicol. Environ. Health*, 1977, *3*, 837-857.

36. Vettorazzi, G.; Radaelli - Benvenuti, B. In International Regulatory Aspects for Pesticide Chemicals, Volume II, Tables and Bibliography; CRC Press: Boca Raton, Florida, 1982, pp 1-9.
37. Joint FAO/WHO Committee on Food Additives. Evaluation of Certain Veterinary Drug Residues in Food, Technical Report Series 788, WHO: Geneva, 1989.
38. Joint FAO/WHO Committee on Food Additives. Evaluation of Certain Veterinary Drug Residues in Food, Technical Report Series 763, WHO: Geneva, 1988.
39. Foxcroft, G. R.; Hess, D. L. In Drug Residues in Animals, Rico, A. G. Ed; Academic Press: London, 1986 pp 147-174.
40. Barry, C. L.; Mac Eachern, G. M. J. Assoc. Off. Anal. Chem, 1983, 66, 4-7.
41. Grandi, A.; Piastrelli, G. Rass. Chim., 1977, 29, 113-120.
42. Gilliam, M.; Argauer, R. J. Environ. Entomol., 1981, 10, 479-482
43. Gilliam, M.; Taber, S. J. Invert. Pathol., 1978, 31, 128-130.
44. Gilliam, M.; Taber, S.; Argauer, R. F. Am. Bee J., 1979, 118, 722-723.
45. Gilliam, M.; Taber, S.; Argauer, R.J. J.Apic. Res., 1979, 18, 208-211
46. EuroResidues. In Conference on Veterinary Drugs in Food. Workshop Reports. Noordwijkerhout, Netherlands, May 1990
47. Burgat-Sacaze, V.; Rico A.; Panisset, J-C. In Drug Residues in Animals, Rico, A.G., Ed.; Academic Press: London, 1986; pp 2-31.
48. MacDonald, A.; Laurencot, H. In Residues of Veterinary Drugs in Food; Haagsma, N.; Ruiter, A.; Czedik-Eysenberg, P.B., Eds.; Proceedings of the EuoResidue Conference, Noordwijkerhout, The Netherlands, May 1990, pp 259-266.
49. Gallo-Torres, H. E. J. Toxicol. Environ. Health, 1977, 2, 827-845.
50. Jaglan, P.S.; Glen, M.W.; Neff, A.W. J. Toxicol. Environ. Health, 1977, 2, 815-826.
51. Lu, A.Y.H.; Miwa, G.T.; Wislocki, P.G. Rev. Biochem. Toxicol., 1988, 9, 1-27.
52. Vroomen, L.H.M.; Berghmans, M.C.J.; vanBladeren, P.J.; Groken, J. P.; Wissink, A.C.J.; Kuiper, H.A. In Veterinary Pharmacology, Toxicology and Therapy in Food Producing Animals, Simon, F.; Lees, P.; Semjen, G., Eds; Unipharma: Budapest, 1990, 259-266.
53. Weber, N.E.; Brynes, S.D. In Residues of Veterinary Drugs in Food; Haagsma, N.; Ruita, A.; Czedik-Eysenberg, P.B., Eds., Proceedings of the EuroResidue Conference, Noordwijkerhout, The Netherlands, May 1990,. pp 404-410.
54. Walton, J.R. Vet. Rec., 1988, 122, 249-251
55. Anon The Public Health Aspects of the Use of Antibiotics in Food and Feedstuffs, Technical Report Series 260, WHO: Geneva; 1963.
56. Greenwood, D. Pig Vet. J., 1990, 24, 38-46.
57. Washington, J.A. Rev. Infect. Dis., 1979, 1, 781-785.

58. Jackson, G. Vet. Rec., 1981, 108, 325-328 and references therein.
59. Somogyi, A. J. Vet. Med., 1989, Suppl. 42, 10-17.
60. Linton, A. H. Vet. Rec., 1981, 108, 328-331.
61. Institute of Medicine. Human Health Risks with the Subtherapeutic Use of Penicillin or Tetracyclines in Animal Feed, National Academy Press: Washington D.C. 1989.
62. Raynaud, J. P.; Mackinnon, J.; Kemp, G.; Beukers, B.; Metzger, K. J. Vet. Med., 1989 Suppl. 42, 40-55.
63. Tancrede, C.; Barakat R. J. Vet. Med., 1989, Suppl. 42, 35-39.
64. Ecobichon, D. J. Pesticides and Neurological Diseases. Ecobichon, D. J; Joy, R.M. Eds.; CRC Press: Roca Baton, Florida, pp 151-203.
65. Bartle, H. New Scientist, 18 May 1991, 30-35.
66. Burns, J. Farmers Weekly, 3 June 1991.
67. Swanston, D.; Shaw, I. Sheep Farmer, October 1990, 31-33.
68. Brown, S.K.; Ames, R.G.; Mengle, D.C. Arch. Environ. Health, 1989, 44, 34-39.
69. Rodnitzsky, R.L.; Levin, H.S.; Mick, D.L. Arch. Environ. Health, 1975, 30, 98-103.
70. Savage, E.P.; Keef, T.J.; Mounce, L.M.; Heaton, R.K.; Lewis, J.A.; Burcar, P.J. Arch. Environ. Health, 1988, 43, 38-44.
71. Duffy, F.H.; Burchfield, J.L.; Bartels, P.H.; Gaon, M; Sim, V.M. Toxicol. Apppl. Pharmacol., 1979, 47, 161-176.
72. Woodward, K.N.; Gray, A.K. Human Exp. Toxicol., 1990, 9, 326
73. Woodward, K.N; Cameron, R.S.; Peirce, M.A. Vet. Rec., 1991, 128, 23-24.
74. Mrvos, R.; Dean, B.S.; Krenzelok, E.P. Clin. Toxicol., 1987, 25, 297-304.
75. Anon, Brit. Med. J. 1987, 294, 652
76. Woodward, K.N. Vet. Human Toxicol., 1991, 33, 168-172

RECEIVED February 28, 1992

Chapter 2

Use of Xenobiotics in Food-Producing Animals in the United States

Regulatory Aspects

N. E. Weber

Division of Chemistry, Center for Veterinary Medicine, Food and Drug Administration, Rockville, MD 20855

A FEDERAL REGISTER document published December 31, 1987 permits the use of carcinogenic animal drugs in food animals as allowed by the DES proviso to the Delaney Clause found in the Food, Drug and Cosmetic Act. A set of guidelines was developed to support the regulation. This chapter will explain the significant toxicology and chemistry elements of those guidelines including metabolic and kinetic aspects that are employed in the regulation of carcinogenic as well as non-carcinogenic animal drugs and feed additives.

This report is designed to give the reader an overview of the regulation of veterinary drugs in the United States. Particular emphasis will be given to the human food safety aspects after the other parts of the regulatory process are outlined so that a broadened perspective of the regulatory scheme may be seen. A profile of regulation of veterinary drugs includes not only the awareness of the need to demonstrate the efficacy of the drug and human food safety aspects but also the aspects of target animal safety, environmental safety and manufacturing controls.

Since the animal drug amendments of 1968, veterinary drugs have had their own section of the Code of Federal Regulations (CFR), section 512. Under this section the sponsor of a new animal drug is required to demonstrate not only the efficacy of the drug, but also the safety to the target animal. These studies must be scientifically valid and well controlled to support the approval and are codified in 21 CFR 514.

Before continuing on to the human food safety and environmental portions, mention should be made that there is a large amount of information required concerning the manufacture and controls for production of the drug prior to its approval. These requirements are very similar if not identical to those for human drugs. The manufacturer must comply with good manufacturing practices (GMPs) and have acceptable stability tests for the product both prior to as well as after the approval.

Environmental Issues

Under the National Environmental Policy Act (NEPA) of 1969, the Center for Veterinary Medicine must consider the potential environmental impact of the actions (decisions) that it takes. The types of action that most commonly require an environmental evaluation are authorizations to investigate uses of new animal drugs and approvals of new animal drugs for marketing. The environmental evaluation of the impact of these actions (and most others) can take one of three forms: (i) a categorical exclusion from preparing an environmental assessment, (ii) an environmental assessment (EA) and (iii) an environmental impact statement (EIS).

The procedures and reports needed to address the environmental issues are described in Title 21 Part 25 of the Code of Federal Regulations (CFR). There is a significant amount of documentation needed to support each of the forms identified above. That documentation usually consists of describing the specific action, describing any controls used to limit the release of the animal drug into the environment and, as necessary, providing information concerning the potential fate and effects of the animal drug in the environment. Often information from the drug metabolism studies used to determine food residue chemistry as well as results of toxicology studies used to determine human food safety can be used as part of the fate and effects information needed to evaluate the potential environmental impact of a new animal drug. The latter also includes safety to the persons who handle and administer as well as produce the drug. The Environmental Sciences Staff at the Center for Veterinary Medicine, FDA, reviews the information submitted.

Human Food Safety

The remainder of this article will outline the major parts of the human food safety portion of the new animal drug approval process with emphasis on the important interface between drug toxicology and residue chemistry. However, before the discussion of the guideline material a brief discussion of some important historical aspects and definitions needs to be given.

Residue. To begin, the definition of a residue has been around for a long time. One of the definitions of a residue comes from the 1958 food additive amendments of the Federal Food, Drug, and Cosmetic Act (FFD&CA), As Amended (*1*), Sec. 409. This section addresses residues from the standpoint of methods as well as safety as follows:

(b)(2)(D) a description of practicable methods for determining the quantity of such additive in or on food and any substance formed in or on food, because of its use;

(c)(5)(A) the probable consumption of the additive and of any substance formed in or on food because of the use of the additive;

Time has not significantly changed the definition of a residue. A recent definition that was given in a FDA 1985 proposed regulation (*2*) is:

"Residue" means any compound present in edible tissues of the target animal that results from the use of the sponsored compound, including the sponsored compound, its metabolites, and any other substances formed in or on food because of the sponsored compound's use.

The phrase "drug residues" is used synonymously with residue. The concepts apply whether the residue comes from an animal drug or feed additive. Another way of visualizing drug residues is seen in the following outline.

DRUG RESIDUES

1. Parent compound
2. Metabolites- addition,cleavage, oxidation, reduction of functional groups
3. Conjugates- small molecules (glucuronides, etc.) macromolecules (bound residues)

Origin of No Residue. The 1958 amendments also had another important food safety concept, the anticancer proviso or the Delaney Amendment as it is often referred to. This provision is incorporated in Section 409 (c)(3)(A) and states "That no additive shall be deemed safe if it is found to induce cancer when ingested by man or animal or if it is found, after tests which are appropriate for the evaluation of the safety of food additives, to induce cancer in man or animal."

In 1962, section 409 of the FFD&CA was amended, in which the same section, (c)(3)(A), quoted above now was given additional language that exempted feed additives through the so called DES PROVISO by stating after the above wording "that this proviso shall not apply with respect to the use of a substance as an ingredient of feed for animals which are raised for food production, if the Secretary finds (i) that, under the proposed conditions of use ...such additive will not adversely affect the animals... and (ii) that no residue of the additive will be found (by methods of examination prescribed or approved by the Secretary ...) in any edible portion of such animal after slaughter or in any food yielded or derived from the living animal." When the FFD&C Act was amended in 1968 to include a new section (Section 512) on New Animal Drugs, the wording of the DES Proviso was included in that section, Sec. 512 (d)(1)(H).

The SOM Procedure. The no residue wording of the DES Proviso which states "...no residue by a method prescribed or approved by the Secretary..." became the foundation by which FDA regulates not only carcinogenic but also non-carcinogenic animal drugs. Beginning in 1973 and continuing for a period of 14 years, the FDA attempted to finalize a regulation to implement the concept by which carcinogenic animal drugs could be approved. The final regulation was published on December 31, 1987 (*3*). The title of the rule is "Sponsored Compounds in Food- Producing Animals; Criteria and Procedures for Evaluating the Safety of Carcinogenic Residues; Final Rule." However, it has become known as the Sensitivity of the Method (SOM) document because it is based on the concept that the Secretary prescribes or approves the methods for carcinogens as permitted by the DES Proviso and through a process determines the level (sensitivity/concentration) required for no residue. A scientific principle applied is that once a drug is given to an animal, residues will not deplete to absolute zero. The focal point of the SOM rule became the procedure by which no residue is determined.

The regulation uses a simple approach which basically involves extrapolating cancer data from laboratory animal models (usually mice or rats) from the observed natural or background incidence to a predicted increased incidence of no more than 1 tumor in 1 million test animals as a result of ingesting the sponsored compound. These calculations involve the number of animals with tumors compared to the total number of animals exposed at a dose in their diet over a lifetime. Various mathematical models are discussed for calculating the 1 in 1 million dose. However, the FDA preferred the multistage model (*4*) at the time the final rule was written. The 1 in 1 million dose becomes the permitted concentration that is used to calculate the no residue level required for the method of analysis for residues in food for human consumption. Although this value is involved in the calculation, an additional calculation is needed to take into consideration total residues in the food animal before a specific value for a specific analyte can become the no residue value by which the drug is regulated. In the following paragraphs, the procedure by which this is done will be outlined.

Carcinogenic Residues and General Food Safety - A Unified Concept

The SOM concept became fully integrated into a general food safety concept that the FDA had developed throughout most of the period prior to the publication of the final rule. FDA uses a similar chemistry approach for determining the tolerance for residues for any compound including "no residue" for a carcinogen. The unified concept applies to carcinogens and non-carcinogens. Before the carcinogenicity of a compound can be evaluated, all compounds have to be treated similarly to assess their carcinogenic potential. An initial decision tree approach that all compounds must undergo was outlined in the SOM document. Threshold Assessment is the name given to this process. The process initially involves a structure-activity assessment to determine whether the sponsored compound is a suspect carcinogen. The compound must also be tested in a battery of mutagenicity tests and must tested in subchronic 90 day studies-usually in the rat and dog. The carcinogenic potential of the compound may be suggested by these tests. If any of the tests signal a potential for carcinogenicity, then chronic lifetime studies are required. When a carcinogenic potential is not seen, the level of residue in edible tissues further determines whether a sponsored compound has to undergo chronic studies. Threshold assessment is outlined below to show the interaction of its elements. Categories D and C refer to suspect and non-suspect carcinogens respectively based on structure-activity relationships. Assignment of category A or B results from the outcome of biological tests which include the mutagenesis battery described below and 90 day subchronic studies in laboratory animals. General Food Safety (GFS) is the category into which a compound is placed when it is determined not to be a carcinogen. The calculation of a safe concentration for residues for GFS is discussed later.

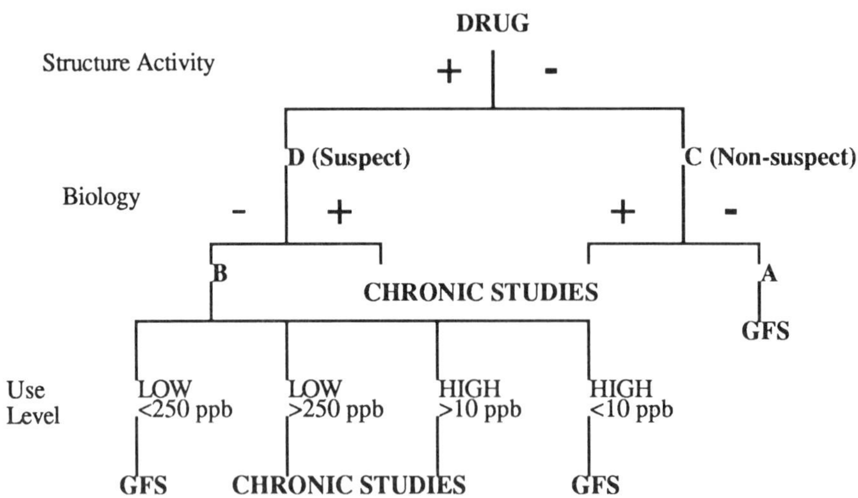

Mutagenicity Testing. A critical part of the threshold assessment is the battery of genetic toxicity and mutagenicity assays. The agency relies greatly on these tests to determine the carcinogenic potential of perspective animal drugs for food animals. The currently used tests are:

1. Bacterial point mutation assay such as the Ames assay with and without S9 activation

2. Mammalian point mutation assay using mouse lymphoma, Chinese hamster ovary or hamster V-79 cells

   or

   Chromosomal aberrations (CAbs), *in vitro* or *in vivo*

3. Unscheduled DNA repair in mammalian cells in culture

Chronic Bioassays. If the result of the threshold assessment supports a decision that chronic bioassays are required to resolve questions involving the carcinogenic potential of the sponsored compound, then long-term bioassays are conducted to assess that potential. The carcinogenicity studies are conducted in the event that toxicology studies along with structure-activity analysis, mutagenicity testing, and allowable residue levels suggest that the compound is a potential carcinogen or must be tested because residue levels are too high. Chronic carcinogenicity bioassay (feeding) studies generally employ two rodent species (rats and mice). At least 50 animals/ dose/ sex are typically used in a three dose plus control group experiment. Dosing is continued until one group of a given sex reaches 20% survival, but not to exceed 30 months duration.

Other Toxicology Requirements. Requirements not previously mentioned but still needed because of special toxicological concerns are a teratology study, a multi-generation reproduction study and a 90 day feeding study in a non-rodent species such as the dog.

Calculation of Safe Concentration. If the compound is not a carcinogen, a no-observed-effect-level or **NOEL** will be determined from non-carcinogenic toxicity end points. The NOEL used in calculating permitted levels for residues is a level of sponsored compound in the diet of the toxicity test species for the most sensitive end point (lowest level) where there was no observed effect. This level is then used in a calculation of the safe concentration (S.C.) for the compound. The calculation also uses safety and food factors as well as a scale up factor for the body weight of man (60 kg):

$$\text{S.C.} = \frac{\text{NOEL(mg/kg in lab species) X 60 kg}}{\text{SAFETY FACTOR X 0.5 kg (meat in diet)}}$$

The safety factor is 100 or 1000 depending on the length of the study. The food factor is 0.5 kg and represents 1/3 of the diet as muscle meat. The permitted amounts of residue in other edible tissues are then considered on the basis on a conservative portion of the diet in which the non-muscle tissue may be consumed. At present the Center for Veterinary Medicine at FDA uses the following consumption factors as multipliers of the safe concentration calculated above (which is for muscle meat): milk =1/3; eggs = 1; liver = 2 to 5 depending on the species; kidney = 3 to 5 depending on the species; and fat = 2 to 5 depending on the species. Poultry kidney is not given a factor because its consumption is insignificant in the diet of people and it is usually removed with the vicera.

If a compound gives a carcinogenic response in the bioassay, the multistage model is used to determine the level of insignificant risk which is considered to be 1 in 1 million (*4*). The mathematically derived value is called the S_o or virtually safe dose. The S_o value is multiplied by consumption factors as described above after it is multiplied by 3 to convert the value to the meat portion of man's diet.

The S_O or the S.C. are permitted concentrations for **total residues** of the animal drug in edible tissues of food-producing animals. They have been calculated from a NOEL in a toxicity study or from an extrapolated 1 in 1 million risk in the diet of animals which exhibited a carcinogenic response in lifetime feeding studies. The total residue concept was discussed earlier and comes from the definition of a residue. All residues that result from administering a feed additive or an animal drug to a food animal are considered as potentially toxic as the parent compound that was fed to laboratory animals unless additional studies are done to remove them from concern.

Chemistry Studies

An outline of how the chemistry portion of animal drugs and feed additives are regulated will now be presented. Several factors are included: (i) total residue data from the food animal; (ii) comparative metabolism data from the laboratory test animal(s) to determine that it has been exposed to all residues of toxicological concern; (iii) calculation of a tolerance for residues in meat, milk or eggs as determined by an acceptable method of analysis and (iv) calculation of a withdrawal time after administration of the drug or feed additive when meat or milk may be marketed for human food. A withdrawal time for eggs is not considered compatible with husbandry practice and is not permitted. The chemistry procedures have been previously outlined (*5*) but must be discussed here so that the entire regulatory concept of the SOM and GFS procedures may be seen.

Total Residue. The first chemistry consideration is to determine the total residue from the feed additive or animal drug when administered to the animal according to label directions. Total residues are usually determined through the use of a ^{14}C radiolabeled drug. Tritium (^{3}H) labeling is sometimes used; however, the sponsor must confirm that the label is in a stable location in the molecule since it is well known that tritium can readily exchange with the protons in water from some locations in most molecules. FDA must also agree that even a ^{14}C labeled molecule is suitably labeled, since some carbon atoms such as those in carboxyl and methyl groups can be metabolized and cleaved from the parent molecule.

Radiolabeled studies are the cornerstone of the chemistry requirements because they fulfill two critical functions. These are: (i) to determine the metabolites of the drug and (ii) to determine the kinetic behavior of the residue as it depletes following cessation of the drug treatment.

A total residue study will ordinarily involve twelve animals which are dosed with the radiolabeled drug according to label directions and are slaughtered at several time points after cessation of treatment. Radioanalysis of tissue samples from the animals will determine the residue depletion kinetics for total residues in the four principal edible tissues (muscle, liver, kidney and fat) and milk and eggs when appropriate. After applying the consumption factors previously mentioned, and plotting the results on a semi-log graph, the last tissue to deplete to its consumption factor adjusted safe concentration is usually selected as the **target tissue**. The target tissue is that tissue in which residues having depleted to their safe concentration will assure the that all of the tissues in the animal are below their permitted safe concentrations.

Metabolism Studies. Having determined the likely choice for a target tissue, the sponsor examines the metabolite profiles during residue depletion to select a **marker residue** that may serve to monitor the total residue during residue depletion in the target tissue. Metabolic profiles should be examined in tissues other than the target tissue to determine that no additional metabolites are present. One of the residues (metabolite or parent compound) in the target tissue is chosen to be the marker residue and its proportion to the total residue is obtained at the point on the total residue depletion curve where this line crosses its permitted safe concentration. The level of the marker residue at that point is called the required level for the marker or R_m. The level is also called the tolerance in the Code of Federal Regulations. Title 21 CFR part 556 contains most of the tolerances for approved animal drugs and feed additives. A graphical presentation of how a R_m or tolerance is established is seen in Figure 1.

Comparative Metabolism in Laboratory Animals. A comparative metabolism study in one of the laboratory test species is also a major part of the human food safety requirements. The objective of this study is to confirm that the toxicology test animal has been exposed to all of the major metabolites found in the food producing species to which people will be exposed in their diets. In this study, rats or mice are fed the radiolabeled drug at doses that they were exposed to during toxicity testing. Dosing for several days is common in order to induce drug metabolizing enzymes. Feces and urine are collected and the radioactive residues in them are profiled to determine whether all of the major metabolites seen in food animals are also produced in the laboratory test animal. Major metabolites are considered as those at concentrations of 0.1 ppm and those making up 10% or more of the observed residues in the tissue of the food animal.

The requirement of this study is to demonstrate that the toxicity test animal has been exposed to all of the major metabolites seen in the food animal. If a major food animal metabolite is not detected in the excreta of the test animal, residues in organs such as the liver and kidney of those animals are then profiled. If major metabolites from the food animal are still not found, then the sponsor must examine the metabolite profiles of the other laboratory species until all of the major food animal metabolites have been accounted for. If a major metabolite from the food producing species is not detected in the test species, then separate feeding studies of the untested major metabolite must be undertaken unless its toxicity can be evaluated by some other means. Chromatographic techniques such as high performance liquid chromatography are often used for profiling metabolites. This type of metabolic evaluation is commonly found in the literature *(6)*.

Analytical Methods for Residues. After the marker residue and target tissue have been identified and a tolerance level or R_m has been set, the sponsor develops determinative and confirmatory methods for the marker residue at the tolerance. The determinative method must be practical and rugged to be useful for routine surveillance monitoring of residues in meat in USDA field laboratories. The confirmatory method is one in which the marker residue is determined unequivocally so that the identity of an above tolerance residue can be supported in a court of law. Methods that are capable of this level of specificity often employ mass spectrometry in one form or another.

Analytical methods which are presented to the FDA undergo a desk review. They are then subjected to a method trial in at least three government laboratories. One USDA and two FDA laboratories are usually engaged to test the methods. In some cases, additional methods for the marker residue are developed after the drug is approved. The marker residue is sometimes included in a screening test by USDA after the drug is approved when broader surveillance by rapid tests is desired by USDA. Screening tests are not required of drug sponsors as a condition of approval at this time. However, the need by USDA to screen large numbers of samples may require the development of a screening test or the inclusion of the marker residue in an existing screening procedure in the future as part of the methods package needed for approval.

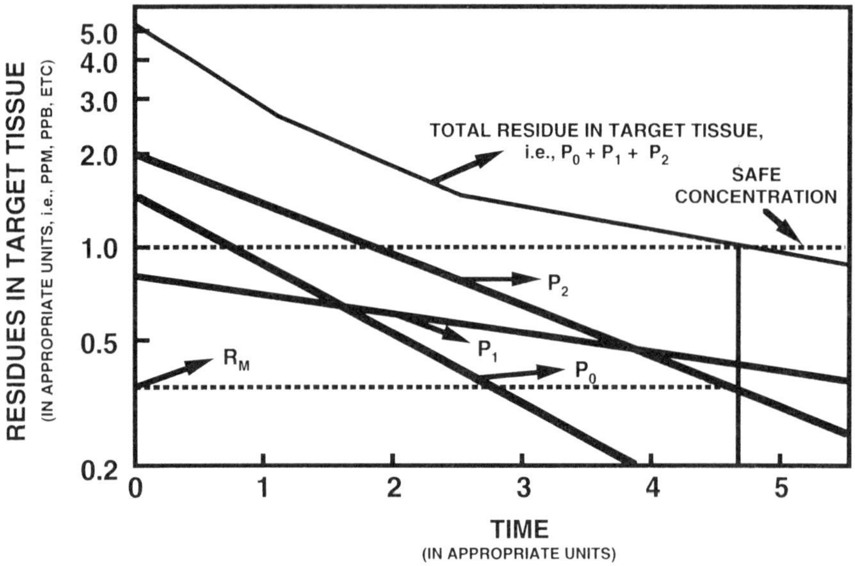

Figure 1. Pharmacokinetic analysis of total residue and metabolite data in a target tissue to determine marker residue and its tolerance (R_m). Reproduced with permission from Dairy Food and Environmental Sanitation **1992**, 12(3), 144-148.

Setting a Withdrawal Time

After acceptable residue methods are developed, the sponsor calculates a withdrawal time for the drug by running a residue depletion study under field use conditions. The objective of this study is to select the times so that at least three are usable, i.e. they lie on the linear portion of the curve running through the tolerance. The food animal is administered the drug for the maximum prescribed conditions of use. In order to obtain an optimal statistical evaluation, 20 animals should be dosed and subsequently slaughtered by normal practice at several intervals after stopping treatment. Four or five animals at each interval and four or five regularly spaced intervals are usually employed. Slaughter interval selection is critical and is best determined by a pilot study. If an acceptable data set is obtained, it is used to determine a withdrawal time by a statistical tolerance limit procedure *(7)* as outlined in agency guideline VI (*3,* implementing guidelines). The 99th percentile tolerance limit with 95% confidence is the one that the agency has chosen. The procedure sets the withdrawal time so that we can be assured with 95% confidence that at least 99% of the animals will deplete to the tolerance within the specified time.

Conclusions

In this article, the procedure established by FDA to regulate food additives and drugs that are used in food producing animals has been reviewed. Although the procedure is complex, the historical basis for the process and the development of the scientific concepts has been outlined. Starting with the definition of a residue and the will of Congress to permit the use of carcinogens in food animals, a regulatory concept for all animal drugs and feed additives has been developed. The SOM procedure begins the process by subjecting a compound to a threshold assessment. Toxicity testing procedures yield a permitted safe concentration or a virtually safe dose in the case of a carcinogen, for total residues in the edible tissues of food animals. Chemistry studies employing the radiolabeled compound determine a target tissue, a marker residue, and a tolerance for the marker residue in meat as well as milk and eggs where appropriate. Methods for determining the marker residue in the target tissue and confirming those residues are developed by the sponsor and evaluated by government laboratories. The methods are subsequently used under field conditions to set a conservative withdrawal time which, if followed, assures that residues are well below a permitted safe concentration. This outline should support the conclusion that residues of approved animal drugs and food additives when used in food-producing animals under approved conditions of use will yield a residue picture that is well within the meaning of safe as determined by a set of scientific principles and societal consensus.

Literature Cited

1. Federal Food, Drug, and Cosmetic Act, As Amended, The Food Additives Amendments of 1958, 72 Stat. 1784, 85th Cong., 2d session.; Sept.6, **1958,** pp. 331-333 (text of the amendment).
2. SOM proposal: Sponsored Compounds in Food Producing Animals; Criteria and Procedures for Evaluating the Safety of Carcinogenic Residues; Proposed Rule and Notice, *Federal Register,* No. 211, Thursday, October 31, **1985,** 50, 45530-45556.
3. SOM final rule: Sponsored Compounds in Food Producing Animals; Criteria and Procedures for Evaluating the Safety of Carcinogenic Residues; Animal Drug Safety Policy; Final Rule, Availability of Guidelines for Human Food Safety Evaluation; Notices, *Federal Register,* No. 251, December 31, **1987,** 52, 49572-49590.
4. Farmer, J.H.; Kodell, R.L.; Gaylor, D.W. *Soc. Risk Anal.* **1982,** 2, 27-34.
5. Weber, N.E. *J. Animal Sci.* **1983,** 56, 244-251.
6. Paulson, G.D.; Feil, V.J. *Drug Metab. Dispo.* **1987,** 15, 841-845.
7. Owen, D.B. *Technometrics* **1968,** 10, 445-478.

RECEIVED February 24, 1992

Chapter 3

Veterinary Medicines: Regulation in Europe and the Importance of Pharmacokinetic Studies

K. N. Woodward

Veterinary Medicines Directorate, Woodham Lane, New Haw, Addlestone, Surrey KT15 3NB, United Kingdom

There is a comprehensive system of legislation in the European Community for the regulation of veterinary drugs which is superimposed onto the regulatory requirements of Member States. An important aspect of this framework is the requirement for studies designed to ensure consumer safety. Pharmacokinetic studies and the data which they generate are important for the understanding of the results of laboratory toxicology studies and in the design and interpretation of residues depletion investigations.

In the European Community (EC), veterinary medicinal products are regulated as two distinct groups. The first of these, and the group which will mainly be featured in the first part of this article constitutes what might be regarded as conventional medicines used for various therapeutic purposes. The second comprises those medicinal products which are added to feed largely for prophylactic, coccidiostatic and growth promotion purposes. The former group is regulated in the EC under a series of Directives, the most important ones being those often referred to as the veterinary medicines directives, 81/851/EEC and 81/852/EEC (1,2). The latter group is controlled under the so-called Feed Additives Directive, 70/524/EEC (3), although it does not cover drugs added to feed for therapeutic purposes; these are dealt with by the veterinary medicines directives. Directives and regulations issuing from the EC are incorporated into the national legislation of Member States. Thus, in the United Kingdom, the requirements of 81/851/EEC, 81/852/EEC and 70/524/EEC have been subsumed into regulations made under the Medicines Act 1968, the key legislative framework for the control of both human and veterinary medicines (4). A key requirement for all veterinary medicinal products under both EC and UK law is that they meet exacting standards of quality, efficacy and safety. Safety here refers to safety to the animal patient, safety to the environment, safety to users and safety to the consumer. This work will

0097–6156/92/0503–0026$06.00/0

examine this latter aspect in terms of what this means in practice set against the framework of EC, and therefore, national requirements. Moreover, it will specifically examine the need for and usefulness of pharmacokinetic studies in safety assessment.

The Veterinary Medicines Directives

These two Directives set out the basic rules for the assessment and authorization of veterinary medicinal products in the EC. Together, they establish the Committee for Veterinary Medicinal Products (CVMP) and the essentials of legislation for multistate licensing of drugs in the Community, as well as providing guidelines governing quality, safety and efficacy. The main role of the CVMP is to advise and give opinions on all matters concerned with veterinary medicines, including those relating to the multistate applications mentioned above. At the time of writing, the predominant method of gaining marketing authorization is through national procedures. A company submits an application to the regulatory authority in the country in which it wishes to market the product. In the case of the United Kingdom, this is the Veterinary Medicines Directorate (VMD) based at Weybridge in Surrey (4). If the application is successful it may market that product but only in the Member State in which the application was made. As suggested in the previous paragraph, the Veterinary Medicines Directives or more precisely, 81/851/EEC, offers an alternative approach - the so-called multistate procedure. Here, a company which has previously obtained an authorization in at least one Member State may request the extension of that authorization to at least five of the others. Under this procedure, a complete dossier of the supporting data is submitted to each Member State and these then have 120 days to assess this and to provide objections to the European Commission. The application and the objections are considered by the CVMP which then issues an opinion which is not binding and Member States then have 30 days to decide whether or not to authorize the product (5).

To assist it in its work, the CVMP has established a number of working groups (6). The Working Group on the Safety of Residues is the EC committee which examines the data relevant to the assessment of consumer safety and which makes recommendations to the CVMP accordingly. One of the major functions of the Working Group on the Safety of Residues is the establishment of maximum residue limits (MRLs) for Community-wide adoption. For the last few years and up to the end of 1991, this has been, and is now a very much *ad hoc* procedure, the priorities for MRL setting being identified by the Working Group. However, from January 1st 1992, a new Council Regulation (EEC No. 2377/90) comes into force (7). This Regulation has two main implications; from January 1992 no Member State may authorize a new pharmacologically active substance for use in veterinary medicines unless a Community MRL has been established and in the period 1992-1996, the safety of all pharmacologically active ingredients currently used in veterinary medicines must be reviewed, and MRLs established. It will be the function of the Working Group to carry out the safety assessment of new and existing active ingredients and to establish MRLs for these.

Directive 87/22/EEC lays down a procedure for applications for so-called high technology products in the Community (8). These are largely products derived from recombinant DNA technology or hybridoma and monoclonal antibody techniques but they also include products given by new delivery systems, and products which feature significant innovations. This procedure is of interest in that it is a semi-centralised procedure. As soon as a Member State receives an application, it must refer it to the CVMP for an opinion. It seems likely that when this involves a new substance, the application will be passed by the CVMP to the Working Group on the Safety of Residues so that an MRL can be established.

Safety Data required

European law with respect to veterinary medicines is somewhat in a state of flux at the present time. The impending implementation of Council Regulation (EEC) 2377/90 has already been referred to but in addition amendments are being made to the veterinary medicines Directives and various guidelines accompanying these are currently being revised. However, the Working Group on the Safety of Residues has produced a draft guidance note which is very much in a final form. This sets out the requirements for pharmacological and toxicological studies in what is referred to as the Safety File, and those for residues and the elaboration of MRLs in the Residues File, for the purposes of 81/851/EEC and Council Regulation (EEC) 2377/90 (9,10)

As described elsewhere in this Volume, the aim of the toxicological (and microbiological) studies is to allow the identification of a no-observed effect level (NOEL) and the calculation of an acceptable daily intake (ADI) (11). The safety file specifies the types of toxicological study that are usually required to allow the construction of the toxicological profile of a veterinary drug active ingredient and to identify the NEL. These include the following:

- single dose toxicity
- repeated dose toxicity
- reproductive toxicity
  - . effects on reproduction
  - . embryo/fetotoxic effects and teratogenicity
- mutagenicity studies
- carcinogenicity studies
- pharmacodynamics in laboratory species
- pharmacokinetics in laboratory species
- microbiological effects
- observations in humans.

The objective of the Residues File is to allow the elaboration of MRLs taking into account the ADI calculated from the toxicology and other safety data, and the pharmacokinetics and residues depletion information. These latter two aspects will now be discussed in more detail.

Pharmacokinetic Requirements: Some Applications and Considerations

Pharmacokinetic studies are often referred to using the acronym ADME - absorption, distribution, metabolism and excretion - reflecting the four main areas for study. The use of such an acronym is simplistic in nature when one considers that each of the four areas is interdependent on one or more of the others, but nevertheless it is a convenient way of dividing up the studies that fall under the broad heading of pharmacokinetics.

The European Communities' guidelines mentioned in the previous paragraph are notable in that while they indicate in some detail the types of toxicological tests required, pharmacology studies appear under the two broad headings of "pharmacodynamic" and "pharmacokinetic" studies. There is a good reason for this for whereas toxicology studies have evolved into rather rigid investigations in order to maintain standards and to ensure that chemicals are tested using similar models, pharmacodynamic, and especially pharmacokinetic studies have remained as *ad hoc* experimental investigations. The design of each experiment depends very much on the nature of the chemical agent and on its interaction (and *vice versa*) with the animal in which the experiments are being carried out. Consequently, each experiment is usually designed *de novo* depending on the drug being tested, the effect being investigated and the animal being used. For these reasons there are no strict pharmacological testing guidelines comparable with those (e.g. the OECD Guidelines) developed for toxicological testing.

For veterinary drugs, intended for use in food producing animals, pharmacokinetic studies are carried out for several reasons and they can be divided into investigations with laboratory animals (mice, rats and dogs for example) and those with target animals (pigs, sheep, cattle etc). In the interpretation of the results of toxicological studies in general, pharmacokinetic studies in laboratory species provide information not only on absorption, distribution, metabolism and excretion *per se*, but they yield in addition much more specific information on the relationship of target organs for distribution with site specific toxicity, data on major metabolites, biochemical mechanisms of toxicity, and they identify routes of excretion, all of which may help in the interpretation of toxicity studies. They also aid in cross species extrapolation of the results of toxicity studies by use of comparative pharmacokinetic data. Measurements of plasma concentrations alone can provide data on the degree of absorption of a drug following administration, its systemic bioavailability, gender, age, and species differences related metabolism and kinetics, profiles after various routes of drug administration, the relationships between dose and systemic exposure with a given dose range, and plasma elimination kinetic data (12-15). These points are recognized in the draft EC Guideline and where necessary, pharmacokinetic findings and those from toxicology studies should be brought together in the Expert Report required by the Directives and by the MRL Regulation.

As well as providing data useful for the assessment of efficacy, pharmacokinetic data in the target animal, are invaluable in identifying target tissues for residues, providing information on the presence or absence of metabolites shown to be toxic, mutagenic or carcinogenic in laboratory species, and perhaps most importantly, providing kinetic data which can be related to residues depletion. From the point of view of residues assessment, these are the most important uses of pharmacokinetic data. An example is provided by the drug carbadox. Carbadox is carcinogenic in laboratory species but when it was evaluated by the Joint FAO/WHO Expert Committee on Food Additives (JECFA) it was noted that the major metabolite present in residues in the pig had been shown to be non-carcinogenic and hence of no risk to consumers (16).

Residues data are time consuming and expensive to produce. Studies involve the treatment of a number of animals with the formulation for which marketing authorisation is being sought, followed by their serial slaughter at various time points after dosing so that residues depletion can be studied. It is often helpful if supporting data from pharmacokinetic studies can be used. One area in which this could prove useful although, in practice it is rarely employed, is in the development of new products following the introduction of an initial formulation. If studies on this initial formulation have correlated pharmacokinetic parameters, especially area under the curve (AUC) measurements, volume of distribution, and plasma and renal clearance rates with residue levels in the depletion study, then residues studies may not be required for new products (for example using higher doses) if pharmacokinetic data for these are available. Moreover, the investigation of pharmacokinetics early on in the development of a drug might not only provide useful data for efficacy considerations, it may also lead to the better design of subsequent residues studies (17-19).

Some caution is, however, required when making comparisons of pharmacokinetic data. The benzimidazole anthelmintics triclabendazole and albendazole respectively show a similar spectrum of metabolites across a range of animal species including humans (21,21). However, in rats and other species, absorption of albendazole after oral administration was in excess of 30% whereas in human it was of the order of 1% (21). With triclabendazole, the C_{max} and AUC values were much higher in ruminants than in horses or humans for the sulphoxide metabolite whereas for the sulphone the horse was more similar to ruminants which differed markedly from the values obtained in man. Sulfamethazine and other sulfonamides show a similar metabolite profile in cattle, sheep and goats but here there are few data on quantitative aspects (22). Consequently it is important to have quantitative pharmacokinetic data in addition to more qualitative information before direct comparisons are made. Qualitative and quantitative differences across species also exist for ivermectin and salicylic acid whereas the aminoglycosides and morantel are essentially similar (23-28).

Age plays an important part in determining the pharmacokinetic behaviour of drugs (29). In sheep for example, the activities of a number of hepatic drug metabolizing enzymes including that of cytochrome P-450, a major component of the microsomal drug metabolizing system, are relatively low in animals aged up to the 6 months when compared with adult levels (30). Such findings probably help to explain why for example the half-lives of sulfadoxine and trimethoprim in neonatal calves and lambs are relatively long, and why they shorten with increasing age (31). It is possible that the effects of gender seen with some drugs are also due in part to the activity of cytochrome P-450, and probably explain the differences in hexobarbital-induced sleeping time in rats where females sleep longer than males (32,33). Antipyrine plasma elimination in rats and cattle shows sex differences and to some extent these differences may be mediated by sex hormones as clearance of antipyrine and sulfamethazine in female dwarf goats was markedly decreased following implantation of trenbolone, a synthetic steroid with anabolic properties (33). These findings have implications for withdrawal periods as residues studies are often conducted in one sex and they underline the need for pharmacokinetic data in their design phase.

Similar sentiments could also be made for the situation involving sick animals. Residues studies are notable in that they are invariably conducted in healthy, mature young animals and yet the drugs being examined are intended for sick animals. Several drugs in several species are known to have different pharmacokinetic behaviour when the animal is sick, a factor which may affect both the efficacy and the withdrawal period. The volume of distribution for orally administered trimethoprim is significantly increased in febrile rabbits compared with their healthy counterparts, and absorption is reduced (34). The prolonged terminal elimination phase of gentamicin is prolonged in the diabetic dog resulting in longer plasma half lives (35). Several drugs including trimethoprim in the calf, sulfamethazine and oxytetracycline in the goat and oxytetracycline in the pig show prolonged plasma elimination half-lives in febrile or infected animals (36-38). On the other hand, the same disease states had no effects on the pharmacokinetics of amoxicillin and chloramphenicol in the calf, nor on ampicillin and sulfamethazole in goats (36,37). Although febrile pigs showed reduced elimination half-lives and increased AUC when oxytetracycline was given orally, there was no effect apparent after intravenous administration suggesting an effect on gastrointestinal absorption (38,39). Such findings lead one to wonder whether residues studies ought to be conducted in sick animals rather than healthy ones or to question whether pharmacokinetic studies should be conducted in both sick and healthy animals so that withdrawal periods might be adjusted where disease states are shown to affect drug clearance.

Finally, it is important to realise that pharmacokinetics may not be the answer to every problem! This is exemplified by reference to residues of sulfamethazine in pigs. Pig kidneys in several countries including the UK and the USA have been found to contain residues of sulfamethazine at above the MRL of 0.1 mg/kg (40-42). The level of violations at one time in the UK (1980-1983) reached around 20% of pig kidneys examined (40) but this has fallen steadily since that period until in 1989 the figure stood at below 6% (43). It might have been tempting at one stage to suggest that the withdrawal periods were of insufficient duration and that pharmacokinetic studies could be useful to throw some light on the problem. However, the real reasons are complex and involve carry over of the drug in the feces and urine of treated pigs, contamination of unmedicated feed with medicated feed, contamination of drinking water at slaughterhouses with feces from treated animals, and failure to observe withdrawal periods (41,44-46).

Residues at the site of injection present specific problems. The persistence of residues at intramuscular injection sites may be due in part to the irritant response produced in the muscle. Chloramphenicol, tylosin, penicillins, dihydrostreptomycin and oxytetracycline have been shown to produce local irritation at the site of injection leading to residues persistence and this may be exacerbated by the solvent used (47-49); with one oxytetracycline product which produced little irritation, residues did not persist (49). Large variations in pharmacokinetic behaviour were noted in addition to the persistence at the injection site and in particular with oxytetracycline, bioavailability was reduced. These studies demonstrate the usefulness of pharmacokinetic data when studying specific routes of administration, and in particular they demonstrate the need to take into account other biological phenomena when attributing withdrawal periods, in this case, irritation at the injection site. The new draft EC Guideline requires that injection sites are examined in residues studies with injectable products and in the case of persistence at the site, then the withdrawal period will be based on this.

Residues may occur in food of animal origin either as the free parent drug, as free metabolites, as bound parent compound or metabolites, or biologically incorporated into endogenous biochemicals (50). The investigation of the nature of residues can be conveniently considered as an area of pharmacokinetics as it involves the study of metabolism, organ distribution, molecular distribution and clearance. It is important as there is an obvious need to correlate the findings in toxicology studies with the nature of the residues present in food of animal origin. It would of course be impractical to subject all known metabolites of a drug to toxicological testing. Some drugs give rise to numerous metabolites e.g. the anthelmintic agent levamisole affords over 50 in the rat (51). It is normally assumed that the metabolites present represent the sum of the toxicity noted. However, bound residues present a particular problem and if they are significant their affect on the ADI must be assessed. JECFA has proposed an approach for the assessment of the bioavailability of bound residues which makes use of extraction procedures and *in vivo* models (16). The *in vivo* systems make use of relay (residue transfer) methodologies whereby the radiolabelled drug is given to the target animal and its tissues containing the bound residues are then incorporated into the food of experimental animals, usually the rat. Bioavailability can then be determined by measuring the radioactivity and distribution in the tissues of the experimental species (52-60). These types of studies have been used to demonstrate that bound residues of ronidazole, carbadox, cambendazole, and furazolidone are of no toxicological concern (61-68). The "bound" residues associated with the milk production enhancer actaplanin and the anthelmintic drug p-toluoyl chloride phenylhydrazone were shown to be moieties metabolically incorporated into normal cellular constituents and therefore of no biological significance (69,70). On the other hand, reactive metabolites can be regenerated from bound residues of trenbolone by hepatic monooxygenases *in vitro* (71), although the toxicological significance for the *in vivo* situation is unclear. Such findings underline the importance of investigations into bound residues, particularly if long withdrawal periods imposed because of long depletion times for total residues as measured by radiolabelling techniques, are to be avoided.

Summary

European Community Directives governing veterinary medicinal products have requirements for pharmacokinetic studies which can be useful in the interpretation of the results of toxicology studies and in the design of residues depletion experiments. There are no strict guidelines for the conduct of these pharmacokinetic studies since design will depend on the nature of the drug under investigation and on the preliminary findings from the toxicology. Used wisely, the results from investigations into pharmacokinetic behaviour can reveal an insight into species and gender differences in toxicity and rates of metabolism, disposition and excretion and help to provide a more complete picture of the biological properties of drugs in laboratory and food-producing animals and provide reassurance on the extremely important issue of consumer safety.

Literature Cited

1. Anon. Council Directive of 28 September 1981 on the approximation of the laws of the Member States relating to veterinary medicinal products (81/851/EEC). *Offic. J. Eur. Commun.*, 1981, No L317, 1-15.
2. Anon. Council Directive of 28 September 1981 on the approximation of the laws of the Member States relating to veterinary medicinal products (81/852/EEC). *Offic. J. Eur. Commun.*, 1981, No L317, 16-28.
3. Anon. Council Directive of 23 November 1970 concerning additives in feeding-stuffs. (70/524/EEC). *Offic. J. Eur. Commun.*, No L270; 840-856.
4. Woodward K.N. *Biologist*, 1991, *38*, 105-108, and references therein.

5. Commission of the European Communities. The Rules Governing Medicinal Products in the European Community. Volume V. Veterinary Medicinal Products. Office for Publications of the European Communities: Luxembourg, 1989.

6. Kidd A.R.M. Brit. Inst. Reg Affairs J., 1989, 8, 13-15.

7. Anon. Council Regulation (EEC) No. 2377/980 of 26 June 1990 laying down a Community procedure for the establishment of maximum residue limits of veterinary medicinal products in foodstuffs of animal origin. Offic. J. Eur. Commun. 1990, No. L224, 1-8.

8. Anon. Council Directive of 22 December 1986 on the approximation of national measures relating to the placing on the market of high-technology medicinal products, particularly those derived from biotechnology. Offic. J. Eur. Commun. 1987, No. L215, 38-41.

9. Commission of the European Communities, Note for Guidance, draft. 1991, Brussels.

10. Commission of the European Communities, The Rules Governing Medicinal Products in the European Community. Volume VI. Establishment by the European Community of maximum residue limits (MRLs) for residues of veterinary medicinal products in foodstuffs of animal origin, draft. 1991, Brussels.

11. Woodward K.N. This volume, and references therein .

12. Clarke B.; Smith D.A. Crit. Rev. Toxicol., 1984, 12, 343-385.

13. Davies D. Xenobiotica, 1988, 18, Suppl. 1, 3-7.

14. Scheuplein R.J.; Shoaf S.E.; Brown R.N. Ann. Rev. Toxicol., 1990, 30, 197-218.

15. Humphrey M.J. In: Pharmacokinetic Considerations in the Conduct and Interpretation of Carcinogenicity Tests. Workshop, Royal Postgraduate Medical School,London,September 1990, 1-4.

16. Anon. Evaluation of certain veterinary drug residues in food. Thirty sixth report of the Joint FAO/WHO Expert Committee on Food Additives, Technical Report Series 799, FAO Rome 1990, 45-50.

17. Sundlof S.F.; Craigmill A.C.; Riviere J.E. J.Vet Pharmacol. Therap., 1986, 9, 237-245.

18. Mercer H.D.; Baggot J.D.; Sams R.A. J. Toxicol. Environ. Health, 1977, 2, 787-801.

19. Dittert L.W. J. Toxicol. Environ. Health, 1977, 2, 735-756.

20. Joint FAO/WHO Expert Committee on Food Additives, In: Toxicological Evaluation of Certain Veterinary Drug Residues in Food, International Programme on Chemical Safety, Geneva, 1990, 3-29.

21. Bogan J.A.; Kinabo L.D.B; Strong M.B.; Formica C.; Galier P.; Alvinerie M. In: Veterinary Pharmacology, Toxicology and Therapy in Food Producing Animals. Proceedings of the 4th Congress of the European Association for Veterinary Pharmacology and Toxicology. Lees P; Simon F.; Semjen G. Eds.; Unipharma: Budapest, 1990, pp 159-163.

22. Juskevich J.C. Drug Met. Rev., 1987, 18, 345-362.

23. Chiu S-H.L.; Taub R.; Sestokas E., Lu A.Y.H.; Jacob T.A. Drug. Met. Rev., 1987, 18, 289-302.
24. Chiu S-H.L.; Green M.L.; Baylis F.P.; Eline D.; Rosegay A.; Meriwether H.; Jacob T.A. J. Agric. Food Chem., 1990, 38, 2072-2078.
25. Chiu S-H.L.; Sestokas E.; Taub R.; Green M.L.; Baylis F.Pl; Jacob T.A.; Lu A.Y.H. J. Agric.Food.Chem., 1990, 38, 2079-2085.
26. Short C.R.; Neff-Davis C.A.; Hseih L.C.; Koritz G.D.; Marlborough M.S.; Barker S.A.; Davis L.E. J.Vet Pharmacol. Therap., 1991, 14, 70-77.
27. Brown S.A.; Riviere J.E. J. Vet Pharmacol. Therap., 1991, 14, 1-35.
28. Lynch M.J.; Mosher F.R.; Levesque W.R.; Newby T.J. Drug Met. Rev., 1987, 18, 253-258.
29. Kietzmann M.; Loscher W. Berl. Munch. Tierarztl.. Wschr., 1990, 103, 277-282.
30. Kawalek J.C.; El Said K.R. Am. J. Vet. Res., 1990, 51, 1736-1741.
31. Pashov D. International Seminar. Pharmacokinetics of Veterinary Drugs, Fougeres, France, 1989, Proceedings.
32. Goldstein A.; Aronow L.; Kalman S.M. Principles of Drug Action: The Basis of Pharmacology, 2nd Ed. Wiley International: London, 1974, p.287.
33. Witkamp R.F.; van't Klooster G.A.E.; van Miert A.S.J.P.A.M. In: Haagsma N.; Ruiter A.; Czedik-Eysenberg P.B. Eds. Residues of Veterinary Drugs in Food, Proceedings of the EuroResidue Conference. Noordwijkerhout, Netherlands, 990, pp 415-419.
34. Ladefoged O. Zbl. Vet. Med., 1979, 26, 580-586.
35. Brown S.A.; Nelson R.W.; Scott-Moncrieff C. J. Vet. Pharmacol. Therap., 1991, 14, 90-95.
36. Groothuis D.G.; van Gogh H.; van Miert A.S.J.P.A.M. In: Trends in Veterinary Pharmacology and Toxicology, van Miert A.S.J.P.A.M.; Frens J.; van der Kreek F.W. Eds.; Elsevier Scientific : Oxford, 1980, pp 77-86.
37. Anika S.M.; Nouws J.F.M.; van Gough H.; Nieuwenhuis J..; Vree T.B.; van Miert A.S.J.P.A.M. Res.Vet.Sci., 1986, 46, 386-390.
38. Pijpers A. Doctoral Thesis, University of Utrecht, 1990, pp 69-83.
39. Pijpers A.; Schoevers E.J.; van Gogh H.; van Leengoed L.A.M.G.; Visser I.J.R.; van Miert A.S.J.P.A.M.; Versheijden J.H.M. J. Vet. Pharmacol., Therap., 1990, 13, 320-326.
40. Anon. Anabolic, Anthelmintic and Antimicrobial Agents. The twenty-second report of the Steering Group on Food Surveillance. The Working Party on Veterinary Residues in Animal Products. Food Surveillance Paper No 22. Her Majesty's Stationery Office: London, 1987, pp 19-20.
41. Lawrence K.; Pig. Vet., 1990, 24, 88-93.
42. McKean J.D. Agripractice, 1988, 9, 15-20.
43. Veterinary Medicines Directorate. Annual Report 1989-90. Ministry of Agriculture, Fisheries and Food: London, 1990, pp 19-20, 39.

44. McCaughey W.J.; Campbell J.N.; Elliott C.T. Vet Rec., 1990, 126, 113.
45. McCaughey W.J.; Elliott C.T.; Crooks S.R.H. Vet Rec., 1990, 128, 125-126.
46. McCaughey W.J.; Elliott C.T.; Crooks S.R.H. Vet Rec., 1990, 126, 351-354.
47. Dagorn M.; Guillot P.; Sander P. Vet. Quart., 1990, 12, 166-174.
48. Nouws J.F.M. Ann. Rech. Vet., 1990, 21, Suppl. 1, 145s-150s.
49. Nouws J.F.M.; Smulders A.; Rappalini M. Vet. Quart., 1990, 12, 129-138.
50. Weber N.E. Drug Met. Rev., 1990, 22, 611-615.
51. Joint Expert Committee on Food Additives. In: Toxicological Evaluation of Certain Veterinary Drug Residues in Food. International Programme on Chemical Safety, Geneva, 1991, 75-101 and references therein.
52. Gallo-Torres H.E. J. Toxicol. Environ. Health., 1977, 2, 827-845.
53. Ferrando R.; Truhaut R. C.R. Acad. Sc. Paris., 1972, 275, 179-283.
54. Ferrando R.; Truhaut R. Toxicol. Eur. Res., 1982, IV, 221-228.
55. Craine E.M. J. Toxicol. Environ. Health, 1977, 2, 905-907.
56. Truhaut R.; Ferrando R. Toxicology 1975, 3, 361-368.
57. Boisseau J. Drug Met. Rev., 1990, 22, 685-697.
58. Gallo-Torres H.E. Drug Met. Rev., 1990, 22, 707-751.
59. Yong M.S. Drug Met. Rev., 1990, 22, 753-751.
60. Laurencot H.J. Drug Met. Rev., 1990, 22, 789-802.
61. Lu A.Y.H.; Wislocki P.G.; Chiu S-H. L.; Miwa G.T. Drug Met. Rev., 1987, 18, 363-378.
62. Lu A.Y.H.; Miwa G.T;. Wislocki P.G. Rev. Biochem. Toxicol., 1988, 9, 1-27.
63. Baer J.E.; Jacob T.A.; Wolf F.J. J. Toxicol. Environ. Health, 1977, 2, 895-903.
64. Ferrando R.; Truhaut R; Raynaud J-P.; Spanogue J-P. Toxicology, 1975, 3, 369-398.
65. Vroomen L.H.M.; Berghmans M.C.J.; van Bladeren P.J.; Groten J.P.; Wissink A.C.J.; Kuiper H.A. In : Veterinary Pharmacology, Toxicology and Therapy in Food Producing Animals. Proceedings of the 4th Congress of European Association for Veterinary Pharmacology and Toxicology. Lees P.; Simon F.; Semjen G. Eds.; Unipharma, Budapest, 1990, pp 259-266.
66. Vroomen L.H.M.; Berghmans M.C.J.; van Bladeren P.J.; Groten J.P.; Wissink A.C.J.; Kuiper H.A. Drug Met. Rev., 1990, 22, 663-676.
67. Wislocki P.G.; Lu A.Y.H. Drug Met. Rev., 1990, 22, 649-661.
68. Sved S.; Foster B. Drug Met. Rev., 1990, 22, 849-861.
69. Saunders J.C. Vet. Rec., 1988, 123, 464-467.
70. Jaglan P.S.; Glenn M.W.; Neff A.W. J. Toxicol. Environ. Health., 1977, 2, 815-826.
71. Evrard P.; Maghuin-Rogister G. Food Add. Contam., 1987, 5, 59-65.

RECEIVED February 28, 1992

Chapter 4

Design and Conduct of Studies To Meet Residue Chemistry Requirements

Residue Depletion and Metabolism of Flunixin in Cattle

R. P. Clement, R. D. Simmons, R. J. Christopher, S. F. Charles, C. N. Casciano, C. B. McCullough, J. F. Lamendola, and M. N. Cayen

Schering-Plough Research, Lafayette, NJ 07848

The safe concentration of drug-related residue must be known in order to determine the withdrawal period for a veterinary product. Often the toxicity data is incomplete and an estimate must be made to progress with requisite residue studies. One approach is to conduct a total residue study with sufficiently widely-spaced sacrifice intervals to assess the rate of depletion of total residue over the projected range of probable safe concentrations. A zero-withdrawal sacrifice interval should be included. The target tissue and marker residue are identified and surveillance/confirmatory assays developed. If a major portion of residue is non-extractable (bound) and the marker is undetectable at times when total residue is still significant, a residue bioavailability study may be necessary. To complete the data package, final residue and comparative metabolism studies are conducted. Studies on the metabolism of flunixin in cattle will illustrate this approach.

To register a new animal drug for use in food-producing animals, the sponsor of the compound must demonstrate that drug-related residues in the edible tissues of treated animals constitute no potential hazard when consumed by humans (1). To this end, the sponsor must develop information on the amount, persistence and chemical nature of the drug-derived residue in the edible tissues of treated animals. This information must then be correlated with that on the metabolism of the compound in the laboratory animal species used for toxicity testing. Because of the time required to complete long-term toxicity studies, residue depletion studies should be designed to develop residue data over the projected range of probable safe concentrations. Close communication between the toxicologist and residue chemist is requisite to adequate study design. In the following discussion,

0097–6156/92/0503–0037$06.00/0

this approach will be illustrated for residue depletion studies conducted with flunixin meglumine in cattle. Flunixin meglumine (Banamine; 2[[2-methyl-3-(trifluoromethyl)phenyl]amino]-3-pyridinecarboxylic acid, N-methylglucamine salt) is a non-steroidal anti-inflammatory drug with analgesic and antipyretic properties which is registered or is under development in many countries as a veterinary pharmaceutical for use in cattle, swine, dogs and horses. The current cattle injectable product being developed in the United States is intended for use as adjunctive therapy in the treatment of bovine respiratory disease. The proposed treatment regimen for cattle is 2.2 mg flunixin active per kilogram body weight per day administered intravenously as the N-methylglucamine (NMG) salt once a day for up to three consecutive days.

Total Residue Depletion Study

The goal of the total residue depletion study is to define the time-dependent depletion of total drug-related residue in the edible tissues (muscle, liver, kidney, fat and, where appropriate, milk for cattle) of target animals (species for which the drug is being developed) following administration of the drug. Radiolabeled (preferably ^{14}C) compound of high radiochemical purity (>98%) is utilized to study the tissue distribution and depletion of radioactivity (representing drug-derived residue). At least 12 animals of the correct species, gender and maturity are treated with the drug by the intended route of administration and highest intended dosage regimen. Groups of three animals are sacrificed at specified intervals after the last treatment and tissues collected. Urine and feces are also collected from at least three animals/sex for a sufficiently long period to characterize the major routes of elimination of radioactivity.

Following collection, tissue and excreta samples are analyzed for total radioactive content and the pooled samples (by sex and/or sacrifice interval) are processed and analyzed to determine the metabolic profile of tissue and excreta radioactivity. Following profiling, major tissue and excreta metabolites are isolated and identified.

Proper conduct of the total residue depletion study is important not only because the results define the depletion of total drug-related residue from the edible tissues of treated animals, but also because this data will be utilized to identify the target tissue (the edible tissue selected to monitor total residue - usually the last tissue in which residues deplete to the safe concentration) and marker residue (residue, i.e. drug and/or metabolite(s), selected to monitor the concentration of the total residue in the target tissue). These results will be utilized to establish the relationship between depletion of total residue and the marker residue in the target tissue. Identification of major metabolites in the urine and feces is important because these products must undergo environmental impact assessment. Tissue and excreta profiles will also be used to establish comparative metabolism in the target and laboratory animal species.

If little or no previous data on the metabolism of the compound in the target species is available, it is prudent to consider conducting a probe residue depletion study in three animals sacrificed at three widely spaced intervals following the last dose. For example, one animal might be sacrificed at 12 hours (zero withdrawal) and another each at 72 and 120 hours post final dose (the exact times selected will depend on the tissue clearance of the drug under development). The information obtained will allow one to select more accurately sacrifice intervals for the definitive study which encompass proposed safe concentrations of total drug-related residue. Additionally, the probe study residue data will allow for adjustments in the specific activity of the radiolabeled dose to ensure tissue concentrations of total radioactivity at later sacrifice intervals which are sufficient for metabolite profiling and isolation.

Initial Total Residue Depletion Study with Flunixin in Cattle. In the initial total residue depletion study, ^{14}C-flunixin NMG was administered once daily for two consecutive days by intravenous injection to three lactating cows and three steers at a dose of 2.2 mg/kg/day (based on flunixin free acid). One cow and one steer per time point were sacrificed at 24, 72 and 120 hours after the final dose, and selected tissues, including liver, kidney, muscle and fat, were collected and analyzed for radioactive content. Highest levels of total radioactivity were noted in the liver and kidneys. Average values of radiolabeled residues in the liver at 24, 72 and 120 hours were 530 $\pm$226, 145 $\pm$49 and 85 $\pm$35 ng equivalents /g, respectively. In kidneys, the average values of total radiolabeled residues at 24, 72 and 120 hours were 520 $\pm$212, 115 $\pm$21, and 55 $\pm$7 ng equivalents/g, respectively. Levels of total radioactivity in fat and muscle were below the limit of reliable detection (defined as 30 dpm above the background count rate) at each sacrifice interval. These results indicate that liver is the target tissue.

Metabolites were isolated and identified in the urine, feces, liver and kidneys of animals administered two consecutive daily intravenous doses of 2.2 mg ^{14}C-flunixin active/kg body weight. Approximately 90% of the total administered dose was recovered in the urine and feces within 24 hours after the second dose. Radioactivity was excreted in approximately equal percentages in the urine and feces. Urine and feces were extracted with methanol and the extracts analyzed by HPLC and TLC. Urinary and fecal metabolites were identified by co-chromatography with known synthetic metabolite standards. The major radioactive component in both cow and steer urine had similar chromatographic retention characteristics to unchanged flunixin. Two minor metabolites, with retention times corresponding to the 2′-methylhydroxy-flunixin and 5-hydroxyflunixin metabolite standards, were also detected. The only radioactive component present in both cow and steer feces had similar elution characteristics to 5-hydroxy-flunixin.

To examine the nature of radiolabeled residues in liver and kidney, the tissues from treated animals were homogenized in methanol, extracted, further cleaned up by solid phase extraction, and analyzed by HPLC. Only the 24-hour post final dose kidney and liver samples were analyzed because of the low levels of total radioactivity present in the other tissues. Three compounds, including flunixin and its 5-hydroxy- and 4′-hydroxy- metabolites were identified by co-elution with authentic standards in the liver and kidney extracts from male and female cattle. The structures of flunixin and its metabolites are presented in Figure 1.

Second Total Residue Depletion Study with Flunixin in Cattle. Because flunixin was only administered for two consecutive days in the initial study, a second total residue depletion study was conducted in which the compound was administered for the maximum proposed treatment duration (3 consecutive days). Twelve Hereford crossbred feeder cattle (6 steers and 6 heifers) received three consecutive daily doses of 2.2 mg/kg flunixin meglumine. Three animals (2 males and 1 female or vice versa) were sacrificed at each of four time points (12, 24, 72 and 120 hours post final dose). Animals in the 12- and 24-hour sacrifice groups were given ^{14}C-flunixin with a specific activity of 0.7 mCi/g. Animals in the 72- and 120-hour sacrifice groups received ^{14}C-flunixin with increasing activities of 1.4 mCi/g and 2.7 mCi/g, respectively. Increasing the specific activity of the dose at the later slaughter groups (72 and 120 hours post final dose) assured that there would be adequate radiolabel present in the tissue samples to provide meaningful results.

Samples of liver, kidney, muscle and fat were removed at sacrifice for analysis of radioactivity. Of the tissues examined, highest levels of total radioactivity were found in the liver and kidneys. No radioactivity was detected in any of the fat or muscle samples collected at 12, 24, 72, or 120 hours. For each animal, the level of total radioactivity in the liver was higher than that in the kidney. Total radiolabeled residues in liver and kidneys were 1,656 ±1,143 ng equivalents/g and 1,123 ±590 ng equivalents/g, respectively at 12 hours after the final dose of flunixin. There was a marked decrease in total radioactivity in both the liver and kidney by 24 hours. Liver and kidney samples at 24 hours post final flunixin dose had levels of total radioactivity of 327 ±121 ng equivalents/g and 308 ±68 ng equivalents/g, respectively. At 72 and 120 hours, total radiolabeled residues in liver were 143 ±36 ng equivalents/g and 121 ±25 ng equivalents/g, respectively. In kidney, the levels of total radioactivity at 72 and 120 hours were 107 ±9 ng equivalents/g and 79 ±15 ng equivalents/g, respectively. This data confirms that liver is the target tissue as it is the edible tissue with the greatest amount of total radioactivity persisting for the longest period of time.

Profiling and identification of flunixin and its metabolites in the liver and kidneys of male and female feeder cattle was also carried out during the second total residue depletion study. Initially, liver tissue from cattle treated intravenously with

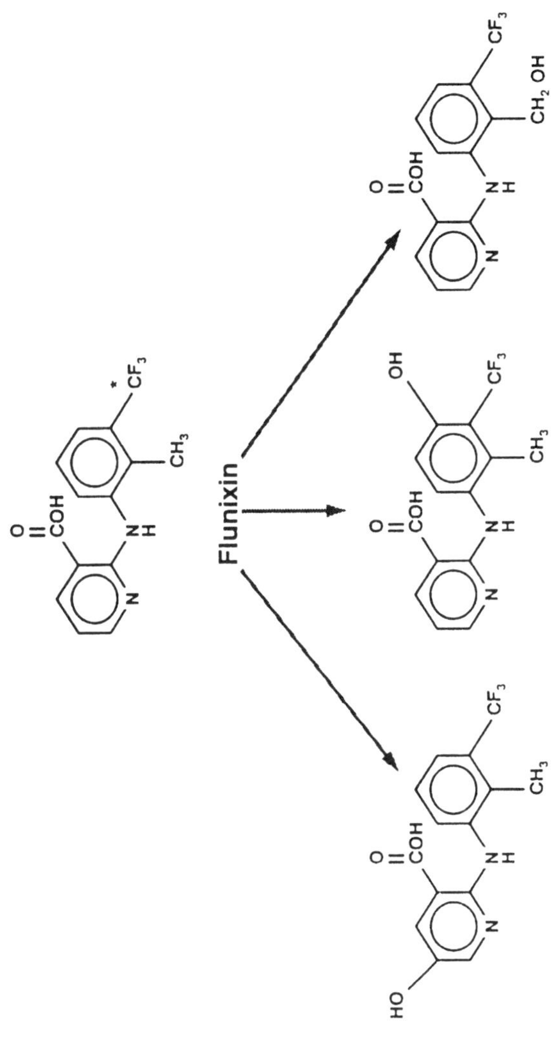

5-Hydroxy-Flunixin 4'-Hydroxy-Flunixin 2'-Hydroxymethyl-Flunixin

Figure 1. Structures of flunixin and its major metabolites (asterisk denotes position of carbon-14 label).

2.2 mg flunixin NMG/kg body weight/day for three consecutive days was homogenized in 60% perchloric acid, incubated at 25°C overnight to digest the tissue, and then extracted with chloroform or methylene chloride under acidic, neutral and basic conditions. The combined organic extracts were evaporated to dryness under reduced pressure, redissolved in methanol and analyzed by HPLC. Recovery of sample radioactivity in the combined organic extracts ranged from 69% to 83% for the 12-hour liver samples and 8% to 22% for the 24-hour samples. Flunixin was the major residue in the organic extract from liver of cattle at 12 and 24 hours post final dose, representing greater than 95% of the extractable radioactivity. In an attempt to enhance recovery, two alternate extraction methods, including aqueous hydrochloric acid extraction and methanol-hydrochloric acid:potassium hydroxide extraction were employed. In each case, recoveries were comparable to the initial results, and flunixin still represented the major residue detected in the 12- and 24-hour liver samples.

Following initial profiling of the 12- and 24-hour post final dose liver tissue, the livers from two of three animals at each sacrifice interval (12, 24, 72 and 120 hours post final dose) as well as kidney tissue from each animal at each sacrifice interval were extracted and the metabolite profiles determined. Each liver and kidney sample was homogenized in hexane to remove lipid, centrifuged, the tissue pellet hydrolyzed in hydrochloric acid, the hydrolyzed homogenate extracted with ethyl acetate at pH 4-4.5, and the ethyl acetate extract concentrated and analyzed by HPLC. With few exceptions, flunixin accounted for at least 50% of extractable tissue radioactivity and was the major residue in the livers and kidneys of male and female feeder cattle. The 4′-hydroxy metabolite of flunixin was also present and represented a major residue in female liver samples at 12 and 24 hours post final dose and in selected male and female kidney samples at 72 and 120 hours. Minor amounts of the other hydroxylated metabolites, including 5-hydroxyflunixin and 2′-methylhydroxyflunixin, were also detected. These results are in agreement with the profiling data from the initial total residue study and indicate that the primary routes of metabolism of flunixin are via oxidation of the pyridine and phenyl ring systems as well as the methyl substituent on the phenyl moiety. The data also indicate that flunixin is the most appropriate marker residue in the target tissue (liver).

Despite use of alternate methods of tissue extraction, the percent of total radiolabeled residue extractable from the liver or kidney decreased with increasing length of time after dosing. In liver, the percent recovery of total radioactivity in the organic extract ranged from 70-94% at 12 hours, 61-72% at 24 hours, and had decreased to between 21 and 48% at 72 and 120 hours post final dose. These results suggest that bound residues may be formed during the metabolism of flunixin in cattle.

Surveillance and Confirmatory Assay Development

To establish a quantitative relationship between the depletion of total residue of toxicological concern and the marker residue in the target tissue, a reliable analytical method for the marker residue must be developed. This method will undergo interlaboratory validation trials and must, therefore, be rugged and of sufficient sensitivity to detect the marker residue at concentrations at which the total residue is at or below the safe concentration. The method usually consists of two components: a surveillance (determinative) assay to quantify the marker residue, and a confirmatory assay to verify the identity of the compound. Each assay must be of acceptable specificity (including freedom from interference by potentially coadministered medications), sensitivity (accurate detection of the marker compound at a concentration equal to one-half the concentration at a time when total residue is at the safe concentration), accuracy and precision. Each method should utilize commercially available reagents and standard analytical laboratory instrumentation, and be capable of being performed by experienced analysts within a 48-hour period. For a detailed discussion of assay requirements, the reader is referred to guideline V of "General Principles for Evaluating the Safety of Compounds Used in Food-Producing Animals" (1).

Assay Development for Flunixin (Marker Residue) in Liver (Target Tissue). A surveillance assay was developed which detects and quantitates flunixin in bovine liver by ion-pair reverse phase high performance liquid chromatography. In the assay, liver samples are homogenized in Tris Buffer (0.1 M, pH 10.7), extracted with dichloromethane/isopropyl alcohol (9:1 v/v) to extract lipids and precipitate proteins, and then, following centrifugation, the aqueous layer is acidified to pH <3 and again extracted with dichloromethane/isopropyl alcohol. The organic layer is separated, concentrated to dryness, and reconstituted in ethyl acetate. The organic layer is washed with a solution of saturated sodium chloride and sodium hydroxide (1N) and then concentrated to dryness, reconstituted in the HPLC mobile phase and analyzed by HPLC, monitoring the effluent at 325 nm. The assay employs clonixin as the internal standard. The assay is linear from approximately 8 to 400 ng flunixin per g tissue with a limit of quantitation (LOQ) of 8 ng/g tissue. The precision was 18% (%CV) or less and the accuracy (relative error) was 12% or less. Mean absolute (vs. nonextracted standard curve) and relative (vs. matrix standard curve) recoveries were 58% and 73%, respectively.

The confirmatory assay for flunixin in bovine liver utilized capillary gas chromatography/mass spectrometry to isolate and confirm the structure of the marker residue. The extraction procedure is similar to that for the surveillance assay, except that following extraction of the acidified aqueous layer, the organic layer is dried with sodium sulfate, concentrated to dryness under nitrogen, derivatized with trimethylsilane, and a portion of the derivatized extract analyzed by capillary gas chromatography/selective ion monitoring-mass spectrometry (GC/SIM-MS).

Following development of the assays for the marker residue (flunixin), liver tissue from feeder cattle dosed intravenously daily for three days with ^{14}C-flunixin NMG (second total residue depletion study) was assayed by the surveillance method. Based on this assay, mean values of flunixin in the liver at 12 and 24 hours post final dose were 531 and 36 ng/g tissue, respectively. Liver samples collected at 72 and 120 hours contained flunixin concentrations below the limit of quantitation of the assay. The flunixin concentrations at 12 and 24 hours represented less than one-third of the total ^{14}C residues, and although no flunixin was detected at 72 and 120 hours, there were still detectable ^{14}C residues at these sacrifice intervals. These results, together with the decreased recovery of extractable total radiolabeled residue noted in liver samples at later time points, suggest the presence of bound residues. Efforts to enhance the recovery of drug-related residue from livers of treated cattle are currently in progress. If these efforts are unsuccessful, and the review of the toxicology data indicates that the safe concentration of total residue is reached at a later time period, then it will be necessary to conduct a relative bioavailability study to assess the percentage of bound residue which is not bioavailable.

Final Residue Depletion Study

The goal of the final residue depletion study is to establish the depletion of the marker residue in the target tissue of treated animals during the last phase of depletion closest to the established tolerance (tolerance refers to the concentration of the marker residue in the target tissue when the concentration of total residue is at or below the safe concentration). It is important to define adequately the marker residue depletion curve because this data will be used to calculate the tolerance limit which, in turn, will define the withdrawal period for the drug. The design of this study should be as close to the actual field conditions of use of the drug as possible. The sponsor should use the final formulation of drug intended to be marketed and it should be administered to animals of correct age and gender at the therapeutic dose and by the proposed route. To establish the residue depletion in tissue, at least 20 animals are treated with the drug by the proposed dosage regimen and then groups of five animals are sacrificed at specified intervals after the last dose and the target tissue collected from each animal. Each tissue sample should be analyzed for the marker residue concentration by the validated surveillance method and the structure confirmed by the validated confirmatory procedure. The data are plotted, statistically analyzed and the tolerance limit calculated as described in the FDA Guideline III (1).

Final Residue Depletion Study with Flunixin in Cattle. In the final residue depletion study, 25 cattle were divided into five groups of three heifers and two castrated males each, and treated with single daily intravenous doses of Banamine solution (final formulation of flunixin NMG) at the rate of 2.2 mg flunixin (active)/kg of body weight/day for three consecutive days. Five animals were sacrificed and liver tissue was collected at 12, 24, 48, 72 and 120 hours after the third dose. Liver samples were analyzed for flunixin by the surveillance (HPLC) and confirmatory (GC-MS) methods.

The mean concentration of flunixin in the liver was 389 ng/g after 12 hours, 53 ng/g after 24 hours, and 13 ng/g after 48 hours of withdrawal from treatment. At 48 hours, only two of the five animals treated had residues above the limit of quantitation (8 ng/g). No flunixin was detected in the 72-hour withdrawal liver samples. Confirmation analyses by GC-MS indicated that detected residues were flunixin. As shown in Figure 2, the flunixin concentrations detected in the livers of treated cattle at 12 and 24 hours post final dose in both the total and final residue depletion studies are similar. However, in both studies, the concentrations of detected flunixin are low with respect to total radiolabeled residue. These results suggest that a significant portion of the total residue may be bound and/or the existing surveillance and confirmatory assays do not adequately extract flunixin residues from the tissue of treated animals. The need for further work on the bound residues and the existing assays will depend on the final safe concentration assigned to the drug.

Metabolism Studies in Laboratory Animals

The goal of metabolism studies conducted in laboratory animals is to demonstrate that metabolism in the laboratory species used for toxicology testing is similar to that in the target animal species. Comparable metabolism in both species indicates that the residues consumed by humans from the tissues of treated target animals have been adequately tested for safety in the toxicology species. The laboratory animal selected for testing is administered repeated daily doses of radiolabeled drug for a sufficient length of time to ensure that the compound has "undergone all relevant metabolic events" (1), including enzyme induction and/or inhibition. Urine and feces are collected during the study and selected tissues, including the tissue designated as the target tissue in the target species, are collected at sacrifice. Tissue and excreta are processed and analyzed to determine the metabolic profile of tissue and excreta radioactivity. To facilitate comparison of the results with those in the target species, the same analytical procedures should be employed.

Metabolism of Flunixin in Laboratory Species (Rat). To study the metabolism of flunixin in laboratory animals, six male and female Sprague-Dawley rats were orally administered ^{14}C-flunixin meglumine by gavage once daily for seven consecutive days at a dose level of 2.5 mg/kg/day (based on free flunixin). The animals were sacrificed at two hours after the final dose and the levels of total radioactivity in selected tissues and body fluids

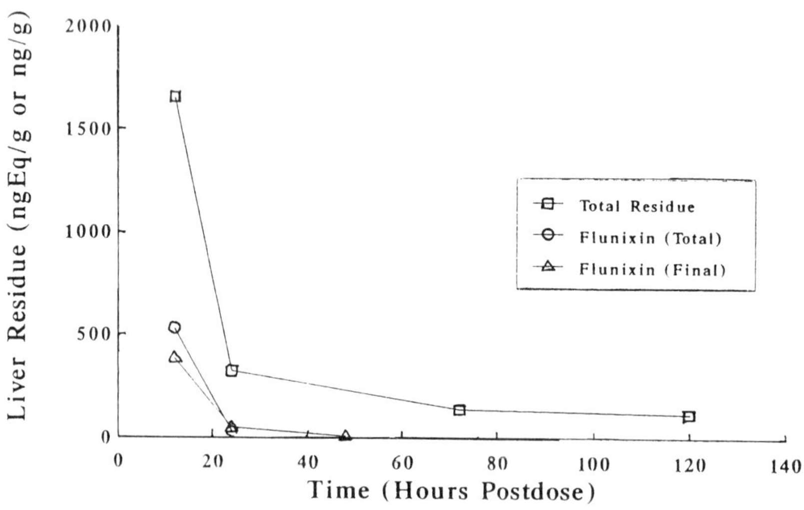

Figure 2. Depletion of total and flunixin residues in livers of cattle following intravenous administration of 2.2 mg flunixin/kg/day for three consecutive days.

were determined. Rat liver, urine and feces samples were profiled for flunixin metabolites by the same analytical methods (HPLC) used to profile cattle tissue and excreta samples in the initial total residue depletion study.

Levels of radioactivity were highest in the liver and kidneys of male and female rats at two hours post final dose and averaged 3.463 µg equivalents/g and 4.073 µg equivalents/g in male and female rat liver, respectively, and 2.383 and 2.132 µg equivalents/g in male and female rat kidney. Urine samples were analyzed by direct injection onto the HPLC column, whereas feces and liver samples were homogenized and extracted with methanol and subjected to solid phase extraction (if necessary) prior to HPLC analysis of the organic extract. The major radiolabeled component in male and female rat urine co-chromatographed with flunixin. Lesser amounts of the two hydroxylated metabolites, 5-hydroxyflunixin and 2′-methylhydroxyflunixin, were also detected. Two major radiolabeled residues, identified by retention characteristics as flunixin and 5-hydroxyflunixin, were observed in rat feces. The major radiolabeled component in male and female rat liver co-chromatographed with authentic flunixin standard. The monohydroxymetabolite, 5-hydroxyflunixin was also detected in male rat liver.

In a second metabolism study, rats were administered a single 10 mg/kg dose of ^{14}C-flunixin by intramuscular injection and urine and feces were collected. Urine samples were profiled by TLC and metabolites identified by co-chromatography with authentic standards. Flunixin was the major component in rat urine following the single intramuscular dose, and lesser amounts of the three hydroxylated metabolites (5-hydroxyflunixin, 2′-methylhydroxyflunixin and 4′-hydroxyflunixin) were also detected. This study illustrates that flunixin is metabolized in a similar manner by rats following either oral or intramuscular administration of the compound.

Results of the rat metabolism studies indicate that, as in cattle, the primary route of metabolism of flunixin is via oxidation of the aromatic ring system(s) and the methyl group on the phenyl ring. Flunixin and the two hydroxylated metabolites, 2′-methylhydroxyflunixin and 5-hydroxyflunixin, were identified in both cattle and rat urine, and 5-hydroxyflunixin was present in the feces of both species. Flunixin and the two hydroxylated metabolites detected in rat and cattle urine were also identified in the liver and kidneys of male and female feeder cattle, and flunixin and the 5-hydroxy metabolite were detected in rat liver. A third hydroxylated metabolite of flunixin, 4′-hydroxyflunixin, detected in bovine liver and kidneys was detected in rat urine following intramuscular administration of flunixin.

Summary

The sponsor of a veterinary product must show that drug-derived residues in the edible tissues of treated animals are safe for consumption by people. In order to do so, data on the amount, persistence and chemical nature of the residues in edible tissues must be determined. In the present paper, the design of studies to generate this information has been discussed. It must be emphasized that each study design (i.e., study protocol) should be reviewed with the regulatory agency (i.e., FDA/CVM) prior to study initiation. Timing of required studies is important, because information derived from one study will be necessary for initiation of other studies. For example, surveillance/ confirmatory assay development cannot be undertaken until the target tissue and marker residue have been identified (in conjunction with the total residue depletion study). Input from toxicology is also important in selecting withdrawal time points for the total and final residue studies. Information on the metabolism of the drug in the laboratory species, if generated early in the development program, can also be useful in the identification of metabolites in the target animal species. Studies on the metabolism of flunixin in rats and cattle, the laboratory and target species, illustrate one approach to acquiring required residue data as well as potential problems encountered (i.e., bound residues).

Literature Cited

1. Center for Veterinary Medicine Guidelines, Number 3, "General Principles for Evaluating the Safety of Compounds Used in Food-Producing Animals" (Guidelines I., through VII.), Food and Drug Administration, (SOM), September, 1986.

RECEIVED March 19, 1992

Chapter 5

Semduramicin in the Chicken

Tissue Residue Depletion Studies

M. J. Lynch[1], G. M. Frame[1,3], J. F. Ericson[1], E. F. Illyes[2], and M. A. Nowakowski[1]

[1]Drug Metabolism Department, Central Research Division, Pfizer Inc., Groton, CT 06339
[2]Animal Health Product Development, Central Research Division, Pfizer Inc., Terre Haute, IN 47808

When biosynthetically prepared ^{14}C-semduramicin sodium, a new anticoccidial ionophore, was fed at 25 ppm in feed to broilers for 7 days, the tissue containing the highest total residues at all withdrawal times was liver. These residues were comprised mainly of unchanged semduramicin sodium (~ 45%) and an array of more polar, low-level (<0.1 ppm) metabolites in liver of poultry sacrificed six hours after withdrawal. Two metabolites were isolated from chicken bile. They were identified by FAB mass spectrometry and confirmed by proton NMR as A-ring O-desmethyl, and G-ring O-desmethyl compounds. Based on a chemical assay procedure utilizing HPLC with post-column derivatization, unchanged residues of semduramicin sodium declined from 0.166 µg/g at 6 hours withdrawal to less than 0.017 µg/g by 24 hours in liver of poultry fed drug in feed at 25 ppm for 44 days.

The use of high intensity rearing systems by the poultry industry has resulted in a dependence on anticoccidial feed additives to provide prophylactic disease control against enteric protozoal infections caused by pathogenic species of *Eimeria (1)*. This dependence on the chemoprophylactic use of anticoccidials in the broiler industry has resulted in a worldwide market demand of $300 million *(2)*.

Among anticoccidials, polyether antibiotics have been the most widely used in the broiler industry over the last two decades because they provide excellent disease control and are refractory to resistance development *(3)*. Polyether antibiotics are branch chained, polyoxygenated carboxylic acids that act as mobile carriers of cations *(4)* by rendering cations lipid-soluble, thereby enabling them to pass across membranes. This process disrupts cationic cross-membrane gradients and is responsible for their anticoccidial activity *(3)*.

[3]Current address: Biological Sciences Laboratory, General Electric Corporate Research and Development Center, Schenectady, NY 12065

0097–6156/92/0503–0049$06.25/0

Semduramicin (Figure 1) is a new potent monocarboxylic polyether antibiotic with antimicrobial and anticoccidial activity *(5)*. It is comparable to several widely used polyether antibiotics *(6)*, including the commercialized anticoccidial feed additives: monesin, salinomycin, lasalocid, narasin and maduramicin. The discovery of semduramicin arose from an effort to identify fermentation-derived products for the treatment of coccidiosis in poultry *(5)*. It is obtained from a fermentation of a mutated strain of *Actinomadura roseorufa (7)*.

The challenge of discovering and developing an agent for the optimal treatment of coccidiosis in broiler flocks involves a wide knowledge of epidemiological, physicochemical, physiological, pharmacological, toxicological, dosing and environmental variables *(8)*. In parallel with elucidation of these considerations, consumer safety requirements must be met.

Studies to define the metabolism and kinetics of drug depletion and to determine safe concentrations of the drug and metabolites in edible tissues of use species are designed to address consumer safety requirements *(9-11)*. Earlier experiments with salinomycin (Lynch, M. J., unpublished data) guided this objective. Also radiotracer metabolism reports for approved polyether antibiotics, including the metabolism of monensin and depletion of its residues in several species (*12*), aided this work. Rapid metabolism and depletion of unchanged monensin was attributed to O-demethylation and oxidation at various positions along the carbon backbone of the molecule; species differences in drug metabolism were generally quantitative. Selective O-demethylation of one of two methoxyl groups has also been described for maduramicin in the chicken and the rat (*13*). O-Demethylation of one or more methoxyl groups and hydroxylation has been reported for residues of maduramicin in turkey excreta (*14*).

Like previous consumer safety studies with polyether antibiotics, the work reported here was undertaken to determine the tissue distribution, metabolic fate and depletion of residues of ^{14}C-semduramicin sodium in broilers and to characterize the depletion of residues of the drug in poultry dosed under commercial use conditions.

Methods and Materials

Instruments. The HPLC system for analysis of unchanged semduramicin was modeled after an LC-post column reaction system described earlier for salinomycin sodium in feeds (*15-16*). The mobile phase consisting of ethyl acetate, iso-octane, glacial acetic acid, triethylamine, methanol (650 + 350 + 4 + 2 + 1) was used in conjunction with a Dupont Zorbax silica 4.6 mm X 25 cm analytical column. NMR spectra were recorded on a Bruker WM-250 spectrometer (modified to incorporate a pulse programmer and Aspect-3000 data system) and a Bruker AM-500 spectrometer, using 50 mg samples dissolved in $CDCl_3$. The ^{13}C and ^{1}H chemical shifts for semduramicin were assigned as reported elsewhere (*17*). FAB (fast atom bombardment) mass spectra were obtained on a VG 70/250 S spectrometer at 1000 resolution.

Labeled Compound. ^{14}C-Semduramicin was prepared with a specific activity of 2.34 mCi/mmole by biosynthesis from sodium [2-^{14}C]acetate in

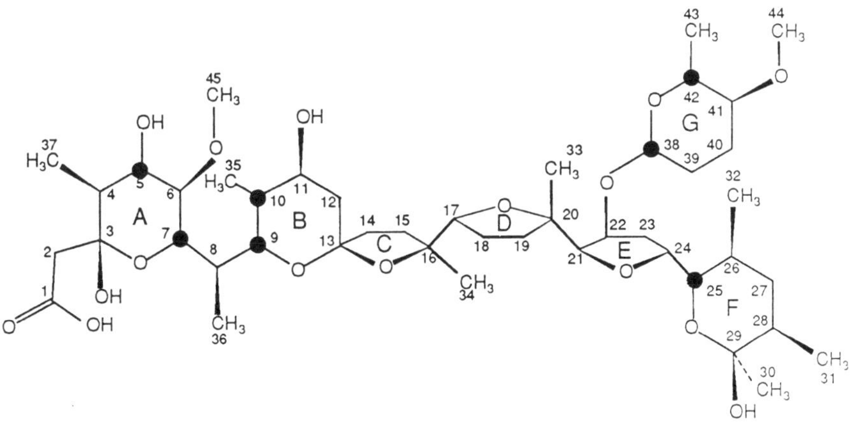

Figure 1. Structure of semduramicin

fermentation with *A. roseorufa* (McArthur, H. A., unpublished data). The incorporation of ^{14}C-label was modeled from a similar fermentation carried out using 99.4% atom purity [2-^{13}C] sodium acetate (New England Nuclear Division of Dupont) as biosynthetic precursor (*18*).

Animal Feeding and Dosing with ^{14}C-Semduramicin Sodium. Fifteen 37-day-old broiler chickens/sex were housed in groups of 7 or 8 per sex in 4 specially constructed large metabolism pens designed for radioactivity studies and of a size sufficient to allow the birds to be as active as those raised in commercial practice. They received *ad libitum* access to non-medicated water and to feed medicated with 25 ppm ^{14}C-semduramicin sodium for 7 days. The average daily consumption of drug was 2.1 mg/kg/day. Three birds/sex were sacrificed 6, 12, 24 and 48 hours after the seventh and final dose day. Samples of plasma, bile, liver, kidney, muscle, fat, and skin with adhering fat were harvested for assay of radioactivity content. Excreta (pooled by sex and homogenized) was collected for each 24-hour period on days 4 through 7 of drug administration.

Animal Feeding and Dosing with Semduramicin Sodium. A group of 48 chicks (24 per sex) was fed *ad libitum* a ration containing 25 ppm semduramicin sodium, the recommended use concentration, for 44 days from the day of hatch. Livers were collected from 3 males and 3 females randomly selected at 6, 12, 18, 24, 36 and 48 hours following withdrawal of the medication. For the identification of metabolites in bile, 35-day-old male broilers were housed in groups of 15 to 24 in 6 pens. They were provided *ad libitum* access to non-medicated water and to feed medicated with 25 ppm semduramicin sodium for 14 days, then sacrificed and bile collected from the gall bladders of 93 birds .

Determination of ^{14}C-Radioactivity in Edible Tissues of Animals. Approximately 1.0 g samples of liver, kidney, and muscle, or 0.3 g samples of fat and skin (with adhering fat) were burned with cellulose powder and Combustaid on a Packard TriCarb Model 306 oxidizer. The burn gases were absorbed in 9 ml of Carbosorb in 20 ml Kimble scintillation vials followed by the addition of 9 ml of Permafluor V scintillator. Samples were counted using a DPM program on a Packard TriCarb 2000CA scintillation analyzer. Aliquots of extraction and HPLC fractions were counted in a Ultima Gold (Packard) scintillation cocktail with a Packard TriCarb 2000CA scintillation analyzer.

Extraction of Residues in Liver or Excreta Homogenates for HPLC Profiling. Homogenized liver or excreta samples from the ^{14}C-semduramicin sodium treated poultry were extracted with methanol:water (8:2) and centrifuged. The supernatant was extracted in sequence with chloroform and ethyl acetate, and the organic phase was concentrated to dryness under reduced vacuum. The residue was then reconstituted in HPLC mobile phase, and analyzed by both liquid scintillation counting (LSC) to determine ^{14}C concentration and HPLC for metabolite profiling. Fractions were collected off the column in 0.5 minute increments over the initial 30 minutes, and then in 1 minute increments thereafter for a total collection period of 1 hour. Each fraction was combined with 15 mL of scintillation cocktail, and the radioactivity was determined with a scintillation counter utilizing a DPM program. The profile of radioactivity in chicken liver was defined by plotting DPM against midpoint time for each HPLC fraction. Percent radioactivity found in each region was based on the total amount of radioactivity recovered from the HPLC corrected for percent recovery through sample preparation. This process provided a means of locating peaks and their relative concentrations.

Extraction of Residues in Bile for HPLC Profiling. Bile recovered from poultry dosed with ^{14}C-semduramicin sodium was diluted with distilled water, extracted with chloroform and the extract was concentrated to dryness. The residue was reconstituted in ethyl acetate/ iso-octane (8:2), and aliquots of this solution were analyzed by HPLC with liquid scintillation counting of fractions to derive the chromatographic profile of radioactivity.

Isolation of Metabolites from Poultry Bile for Spectral Identification. Pooled samples of bile from poultry receiving feed medicated at 25 ppm semduramicin sodium for 7 days were extracted with ethyl acetate followed by chloroform. Abundant metabolites were isolated by collection of fractions from repeated injections on a normal phase silica HPLC column. The recovered fractions were analyzed by Fast Atom Bombardment Mass Spectrometry and proton NMR spectroscopy.

Reverse Isotope Dilution Analysis for Carbon-14 Labeled Residues of Unchanged Semduramicin Sodium in Poultry Liver. Before tissue homogenization nonlabeled semduramicin sodium (5 µg) was added to 1 g of tissue as carrier tissue homogenization, and then extracted with 5 mL of methanol-water (8:2). A supernatant, obtained by centrifugation, was washed with 10 mL of n-hexane. Separation of the phases by centrifugation and aspiration of the extract yielded a polar phase that was extracted with 10 mL of chloroform. The extract was evaporated to dryness under a stream of nitrogen at 55°C, redissolved in chloroform, and passed through a Silica BondElute (American Bioanalytical) using a vacuum manifold (American Bioanalytical). Following the addition of 5 mL of chloroform to the cartridge, semduramicin was eluted with 2.5 mL of methylene chloride: methanol (9:1). For HPLC analysis, the eluate was evaporated under nitrogen at 55°C and redissolved in 150 µl of mobile phase, containing an analog (UK-58,582) as internal standard. The sample was injected onto HPLC containing a vanillin post-column reaction cell, and the fraction eluting with the retention time of semduramicin sodium was collected for scintillation counting.

The total amount of semduramicin sodium recovered from the HPLC was determined by reference to a standard calibration curve that related the peak area ratio of semduramicin sodium to the internal standard. The radioactivity recovered from the HPLC was corrected for the recovery of "cold" semduramicin sodium. In this procedure the dynamic range of the calibration curve for the nonlabeled semduramicin sodium extended from 1-40 µg/mL, and the dynamic range for tissue fortified with ^{14}C-semduramicin sodium ranged from 0.010 to 0.250 µg/g. Control tissue fortified with ^{14}C-semduramicin, sodium at 0.010, 0.025, 0.050, 0.100 and 0.200 µg/g gave mean levels with standard deviations of 0.011 (0.007), 0.025 (0.003), 0.050 (0.003), 0.095 (0.003), and 0.192 (0.011) µg/g. The recovery of "carrier" semduramicin sodium was 54%.

Chemical Assay of Semduramicin in Poultry Liver by HPLC. Semduramicin concentrations were determined in liver by HPLC utilizing post-column derivatization (Ericson, J. F. *et al.*, unpublished data). In this procedure, semduramicin sodium was extracted from liver with methanol-water, separated and concentrated by solid phase extraction steps, and determined by HPLC with post-column derivatization. Sample preparation involves extraction of semduramicin sodium from liver with aqueous methanol followed by separation of the drug from coextractives using reverse-(C_8) and normal-(silica gel) phase extraction columns in sequence. The HPLC procedure relies on a post-column reactor for quantitating semduramicin sodium in purified extracts of poultry liver using vanillin (4-hydroxy-3-methoxybenzaldehyde) as a derivatizing reagent (*15, 16, 19, 20*). The chromophore is formed under anhydrous, acidic conditions in which the vanillin reacts with the hydroxyl groups of semduramicin sodium yielding a derivative that strongly absorbs at 522 nm. Heat is required to drive the reaction, but the required temperature is easily workable within the HPLC system. This type of reaction has been used in conventional colorimetric analysis for monensin (*21*) and for lonomycin (*22*). The ionophore is chromatographed on a normal phase column within a total run time of fifteen minutes. The percent recovery of semduramicin sodium ranged from 70.9% to 118.4% with a mean of 84.3% and a coefficient of variation of ±13.5% over the range of 0.010 to 0.300 µg/g.

Results

The ^{13}C and ^{1}H NMR chemical shifts for semduramicin sodium were assigned (Table I) as reported elsewhere (*17*). These spectral data revealed the following groups: CH_3 (9), CH_2 (10), CH (5), CH_3O (2), O-CH (12), OCHO (1), OCOH (1), CO (2), OCO (2), and COONa (1). A comparison of the ^{13}C-enriched and unenriched ^{13}C-NMR spectra is displayed in Figure 2 and confirmed the significant incorporation of [2-^{13}C]acetate. Using the assignment data in Table I and comparing the two spectra utilizing the ^{13}C spin-spin interaction data from the enriched preparation, carbons were classified in the structure as being enriched by ^{13}C from acetate labeled in the 2 position (Figure 3).

The following conclusions were drawn from the locations of strong enrichment:

a. Isotope enrichment occurs at many sites along the main backbone of cyclic polyethers in the molecule.

b. Little to no enrichment occurs in the sugar substituent (ring G) or in the methoxy carbons.

c. All backbone cyclic ether rings contain at least one enriched carbon.

d. The wide distribution of label ensures that any important metabolite will retain substantial label.

TABLE I. ^{13}C and ^{1}H NMR Chemical Shift Data for Semduramicin Sodium in $CDCl_3$ Including Sites (*) for Incorporation (^{13}C+) of [2-^{13}C]acetate

Carbon	*Functionality*	*^{13}C Shift[a]*	*^{1}H Shift[b]*	*^{13}C (+)*
1	COONa	179.09 (0)	–	
2	CH_2	45.38 (2)	2.17, 2.49	*
3	O-C-O	97.70 (0)	–	*
4	CH	45.28 (1)	1.48	*
5	O-CH	74.74 (1)	3.71	*
6	O-CH	81.95 (1)	3.11	*
7	O-CH	66.80 (1)	3.73	
8	CH	33.74 (1)	1.98	*
9	O-CH	67.61 (1)	4.23	
10	CH	33.54 (1)	1.81	*
11	O-CH	70.03 (1)	3.92	
12	CH_2	33.77 (2)	1.62, 1.90	
13	O-C-O	107.45 (0)	–	*
14	CH_2	38.89 (2)[c]	1.73, 1.97	*
15	CH_2	33.39 (2)[c]	1.76, 1.98	
16	C-O	84.51 (0)	–	*
17	O-CH	82.28 (1)	3.53	*
18	CH_2	26.83 (2)	1.47, 1.71	*
19	CH_2	32.25 (2)	1.50, 2.40	
20	C-O	84.15 (0)	–	*
21	O-CH	86.96 (1)	4.03	*
22	O-CH	80.92 (1)	4.16	*
23	CH_2	32.47 (2)	2.23	
24	O-CH	80.22 (1)	4.49	*
25	O-CH	73.01 (1)	3.93	*
26	CH	33.11 (1)	1.23	*
27	CH_2	36.40 (2)	1.32, 1.42	*
28	CH	39.81 (1)	1.43	*
29	O-CH	96.89 (0)	–	*
30	CH_3	26.05 (3)	1.29	

Continued on next page

TABLE I. *(continued)*

Carbon	*Functionality*	*13C Shift* [a]	*1H Shift* [b]	*13C (+)*
31	CH_3	16.99 (3)	0.91	
32	CH_3	17.51 (3)	0.87	
33	CH_3	23.25 (3)	1.12	
34	CH_3	27.56 (3)	1.49	
35	CH_3	10.43 (3)	0.84	
36	CH_3	11.05 (3)	1.08	
37	CH_3	112.10 (3)	11.03	
38	O-CH-O	103.22 (1)	4.41	
39	CH_2	30.55 (2)	1.53, 1.80	
40	CH_2	26.92 (2)	1.31, 2.18	
41	O-CH	79.83 (1)	2.81	
42	O-CH	74.57 (1)	3.31	
43	CH_3	18.38 (3)	1.24	
44	O-CH_3	56.86 (3)	3.36	
45	O-CH_3	59.03 (3)	3.52	

a In ppm from TMS in $CDCl_3$ solution; number of attached protons in parenthesis.
b In ppm from TMS in $CDCl_3$ solution.
c Tentative assignment.
* Denotes ^{13}C enrichment.

Radiotracer Studies in Poultry. The radiotracer study characterized the depletion of total residues in edible tissues of 37-day old broiler chickens that received 7-day's *ad libitum* access to feed medicated with 25 ppm ^{14}C-semduramicin sodium. Residues in edible tissues were determined at 6, 12, 24, 48 and 120 hours after treatment. The tissue containing the highest total residues at all withdrawal times was liver. Liver residues were 1.8 or more times higher than residues in the next highest tissues, fat or skin. Total residues in liver were depleted from a mean value of 0.273 µg/g at 6 hours to 0.058 µg/g at 24 hours, and eventually to 0.018 µg/g at 120 hours (Table II).

An assessment of the distribution and depletion of residues in the edible tissues of broilers was complemented with a determination of the concentrations of drug residues in plasma, bile and excreta. Mean plasma radioactivity at 6 hours withdrawal was 0.025 µg/mL and declined below the quantitation limit of 0.009 µg/mL by 24 hours. Bile radioactivity ranged

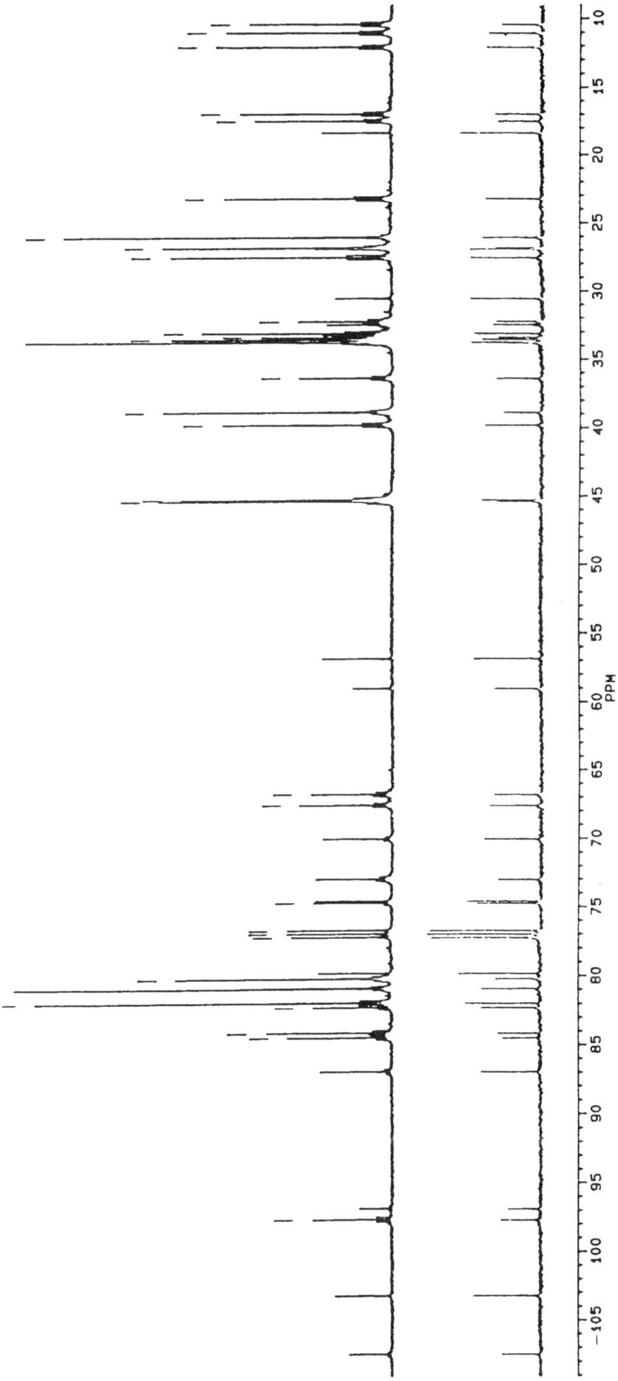

Figure 2. ^{13}C-NMR spectra of enriched and unenriched semduramicin sodium in $CDCl_3$.

from 9 to 50 µg/mL at 6 hours withdrawal and declined to 0.08 µg/ml by 120 hours (Table III). Overall mean radioactivity in pooled excreta ranged from 19.4 to 26.6 µg/g during the last four of the seven days of *ad libitum* administration of feed medicated at 25 ppm (Table IV).

Total residues in large-scale pooled homogenates of the target tissue, liver, from all birds at each withdrawal time were also subjected to assay for unchanged drug concentration by reverse isotope analysis. Radiometric measurement of unchanged drug in the pooled liver homogenates by reverse isotope dilution analysis revealed that semduramicin sodium was depleted in a biphasic pattern similar to that of total residues (Figure 4).

Determination of the Ratio of Unchanged Drug to Total Residues by Chemical Assay. In conjunction with the determination of total residues in poultry liver samples from the definitive radiotracer study, unchanged drug concentrations from the same samples were assayed by a chemical HPLC procedure to measure the fraction of total residues represented by the unchanged drug. This fraction decreased from 0.48 at 6 hours to 0.30 at 12 hours and to less than 0.21 by 24 hours withdrawal (Table V). Thus, monitoring unchanged semduramicin levels in liver by chemical assay can provide a collateral method of assessing total residues of the drug.

Metabolic Profile of Radioactivity in Liver, Bile and Excreta. Based on these observations, liver was chosen to define the metabolite profile of semduramicin sodium in poultry. As defined by HPLC, residues in liver of poultry sacrificed six hours after withdrawal were comprised mainly of unchanged semduramicin sodium (~ 45%) and an array of more polar, low-level (<0.1 ppm) metabolites. None of the radioactive components, labeled "A" through "S" (Figure 5) represented more than 10 percent of total radioactivity. For comparison, the profiles of radioactivity recovered from bile of a female broiler sacrificed 12 hours after withdrawal, and of a 24-hour excreta sample collected from broilers during the seventh day of dosing are given in Figure 5. When compared to the profile in liver, each had relatively higher amounts of metabolites.

Table II. Mean(1SD) Total Residues of ^{14}C-Semduramicin (µg/g equivalents of Na salt of drug) in Poultry (3 sex/time point) Given *ad libitum* Access to Feed Medicated at 25 ppm for 7 Days

Time (Hrs)	*Liver*	*Kidney*	*Muscle*	*Fat*	*Skin/Fat*
6	.273(.079)	.051(.012)	.015(.006)	.074(.018)	.057(.015)
12	.112(.023)	.027(.018)	.007(.003)	.027(.008)	.022(.006)
24	.058(.005)	.012(.001)	.003(.001)	.015(.003)	.015(.003)
48	.031(.005)	.006(.001)	.002(.001)	.011(.002)	.011(.001)
120	.018(.003)	.004(.001)	.001(.000)	.010(.002)	.009(.002)

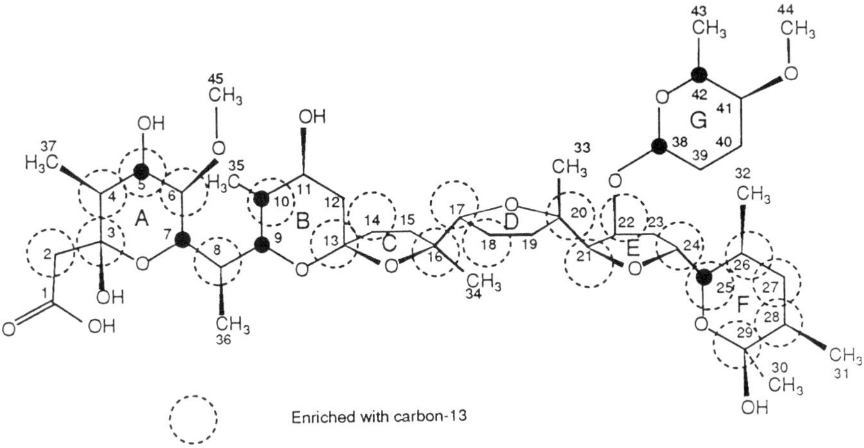

Figure 3. Location of ^{13}C-enrichment of semduramicin from fermentation synthesis with [2-^{13}C]acetic acid.

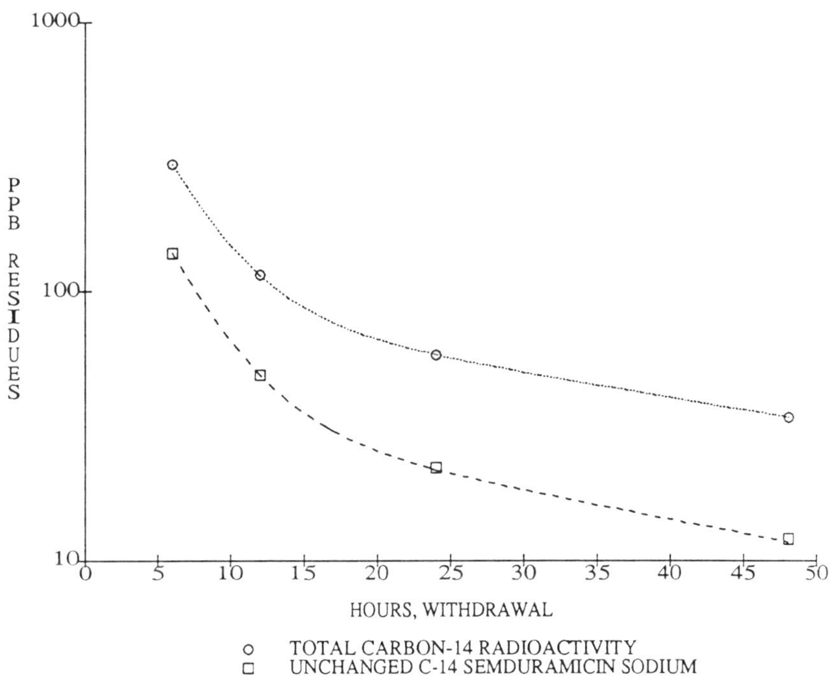

Figure 4. Radiometric determination of the decline of total and unchanged residues in livers of poultry fed carbon-14 labeled semduramicin sodium at 25 ppm for 7 days.

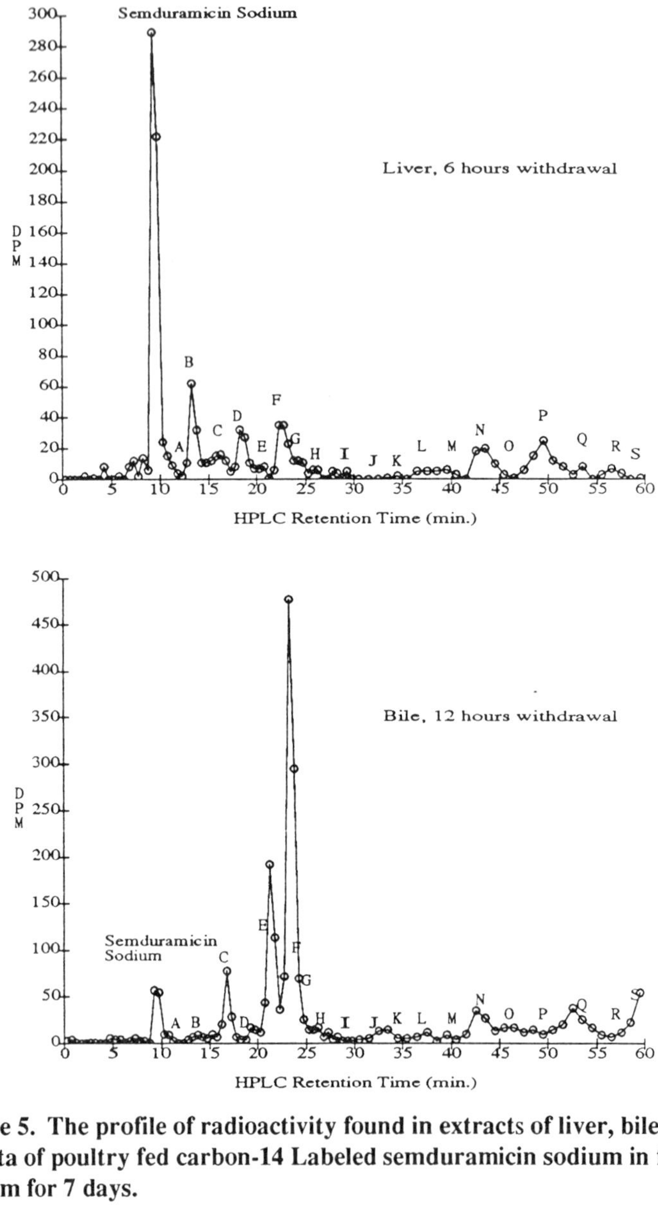

Figure 5. The profile of radioactivity found in extracts of liver, bile and excreta of poultry fed carbon-14 Labeled semduramicin sodium in feed at 25 ppm for 7 days.

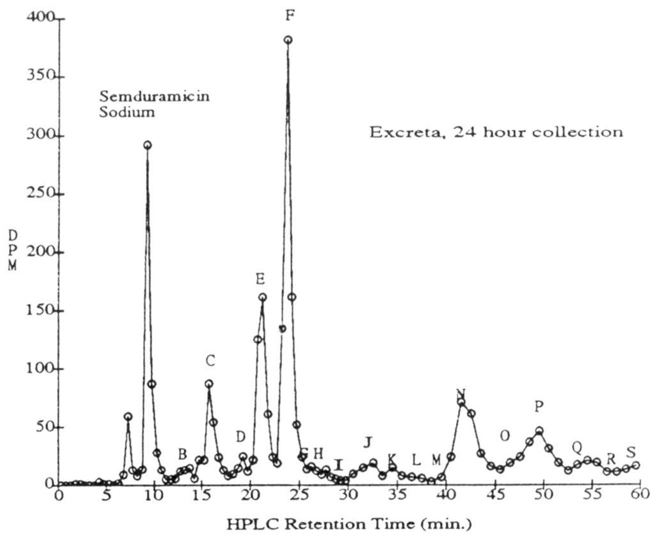

Figure 5. Continued.

Table III. Mean Plasma and Bile Radioactivity (µg/ml of ^{14}C-semduramicin sodium equivalents) in Poultry (3 sex/time point) Given *ad libitum* Access to Feed Medicated at 25 ppm for 7 Days

Withdrawal Time (Hrs)	*Semduramicin ppm Plasma*	*Bile*	*Withdrawal Time (hrs)*	*Semduramicin ppm Plasma*	*Bile*
6	.025	22.60	48	<.009	0.75
12	.010	6.00	120	<.009	0.08
24	<.009	5.27			

Table IV. Pooled Excreta Radioactivity (µg/g of semduramicin sodium equivalents) Collected From Poultry Receiving *ad libitum* Access to Feed Medicated at 25 ppm for 7 Days

24-Hour Collection	*Males*	*Females*	*Mean*
Day 4	29.8	19.1	24.5
Day 5	22.2	16.6	19.4
Day 6	29.8	23.3	26.6
Day 7	18.7	27.0	22.9

Table V. Determination of the Ratio of Unchanged Drug to Total Residues of Carbon-14 Labeled Semduramicin Sodium in Livers of Poultry Fed ^{14}C-semduramicin Sodium in Feed at 25 ppm for 7 Days

Withdrawal Time, Hours	*Total C-14 Residues, µg/g*	*Unchanged Drug Residues, µg/g*	*Ratio of Unchanged Drug to Total Residues*
6	0.273	0.132	0.48
12	0.112	0.035	0.31
24	0.058	<0.012	<0.21

Spectral Identification of Metabolites "E" and "F". Interpretation of the results of mass spectrometry of semduramicin sodium and its metabolites was facilitated by comparison with the fast atom bombardment mass spectral (FAB-MS) analysis of maduramicin sodium and other ionophores (*23*), the thermospray liquid chromatography/tandem mass spectrometry of maduramicin and its metabolites (*14, 24*), and the extensive review of the mass spectrometry of polyether antibiotics (*25*). In this way, mass spectral interpretations were formulated for semduramicin and metabolites designated "E" and "F". Mass fragments observed for semduramicin and two metabolites recovered from bile are compared in Table VI. Similarities in the fragmentation patterns of semduramicin sodium and metabolites "E and "F" strongly suggested closely related compounds with a difference of 14 daltons separating the metabolites from unchanged drug. A decrease of 14 daltons between semduramicin sodium and its two metabolites is consistent with O-demethylation of the "A" and "G" ring methoxy groups. Possible structures for metabolites "E" and "F" are given in Figures 6 and 7. Assignment of these structures was confirmed by ^{1}H-NMR.

Table VI. FAB MS Data for Semduramicin Sodium and Metabolites "E" and "F"

	FAB MS Fragment Ions		
Proposed Ion Assignments	*Unchanged Semduramicin*	*Metabolite "F"*	*Metabolite "E"*
M+Na	895	881	881
M+Na - H_2O	877	863	863
M+Na - CO_2	851	837	837
M+Na - H_2O - CO_2	833	819	819
M+Na - 2(H_2O) - CO_2	815	801	801
M+Na - H_2O - CO_2 - CH_3OH	801	787	787
M+Na - H_2O - CO_2 - $C_6H_{12}O$	733	719	719
M+Na - H_2O - CO_2 - $C_7H_{12}O_2$	705	-	691
M+Na - $C_{18}H_{30}O_9$	505	-	491
M+Na - $C_{17}H_{28}O_9$	-	505	-
M+Na - $C_{22}H_{36}O_{10}$	-	421	-
M+Na - $C_{23}H_{38}O_{10}$	421	-	-

Confirmation of the Identities of Metabolites "E" and "F" by ^{1}H-NMR. Metabolites "E" and "F" of semduramicin were isolated in sufficient quantity to obtain ^{1}H-NMR spectra and to characterize structural changes based on observed differences with the parent structure. Although the most intense signals were from "experimental background", the spectral intensity of the two methoxy methyl groups was sufficiently clear to permit unambiguous assignment.

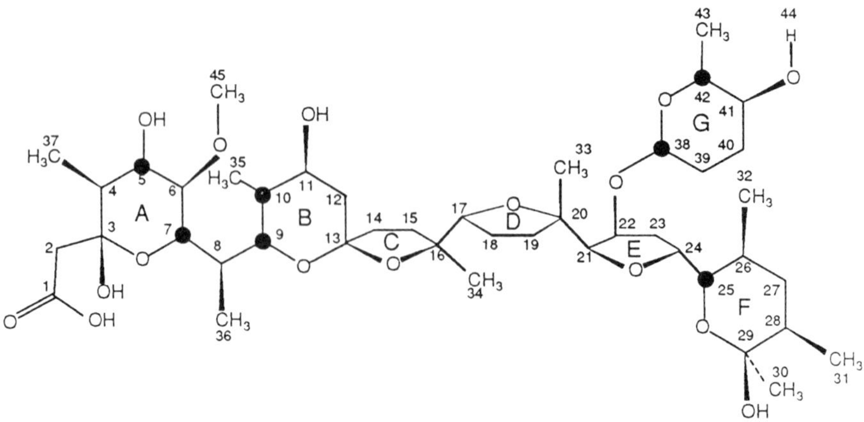

Figure 6. Semduramicin metabolite "E"

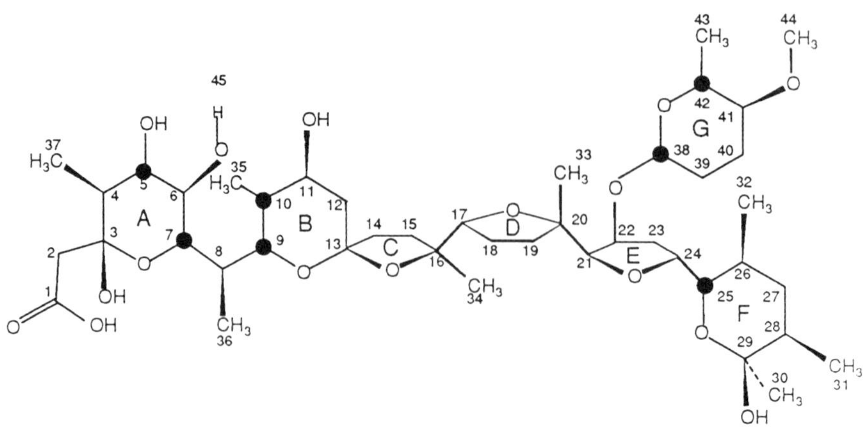

Figure 7. Semduramicin metabolite "F"

As illustrated in Figure 8, the ^{1}H-NMR spectrum of semduramicin sodium showed two sharp singlets for the two methoxy groups, 3.52 ppm for the methyl group labeled at position 45 in the "A" ring and 3.36 ppm for the methyl group labeled as position 44 of the "G" ring (Figures 6 and 7).

Analysis of the ^{1}H-NMR spectrum of metabolite "E" showed only a singlet at 3.52 ppm suggesting loss of the ether linked methyl at position 44. This was corroborated by observed changes in the chemical shift for protons in positions 41 and 42, but none for the proton in position 6; it remained unchanged relative to semduramicin (apparent triplet at 3.11 ppm).

Likewise, analysis of the spectrum of metabolite "F" showed the methyl singlet at 3.36 ppm intact suggesting loss of the ether linked methyl at position 45. Loss of the apparent triplet at 3.11 ppm (proton in position 6 of semduramicin) and the presence of a seemingly unchanged multiplet at 2.81 ppm (similar to proton in position 41 of semduramicin) supported this interpretation. Thus, the structures postulated for metabolites "E" and "F" by mass spectrometric analysis were confirmed by ^{1}H NMR.

Semduramicin Residues in Poultry Fed Under Proposed Use Conditions. Following the evaluation of the decline of total and unchanged residues in poultry dosed with radiolabeled drug, unchanged semduramicin sodium concentrations were determined in liver specimens of poultry fed nonlabeled semduramicin sodium in feed at 25 ppm for 44 days and withdrawn for 6, 12, 18, 24, 36 and 48 hours. Analyses of purified extracts of liver by normal-phase HPLC with post-column vanillin derivatization detection revealed unchanged semduramicin sodium depleted from an initial level of 0.166 µg/g at 6 hours to less than 0.017 µg/g by 24 hours withdrawal (Table VII). Residues at the 36 and 48 hour withdrawal periods were less than 0.010 µg/g. These results were consistent with the decline in unchanged drug residues determined in the radiotracer study.

Table VII. Mean (1SD) Residues (µg/g) of Semduramicin Sodium in Liver of Poultry (3 sex/time point) Given *ad libitum* Access to Feed Medicated at 25 ppm for 44 Days

Hours	*µg/g* *	*Hours*	*µg/g*
6	0.166 (.104)	18	0.036 (.045)
12	0.044 (.020)	24	<0.017

* The lowest limit of detection was 0.010 µg/g.

Discussion

These experimental results indicate that the biosynthetic incorporation of [2-^{14}C]acetate for preparing labeled semduramicin sodium, and the

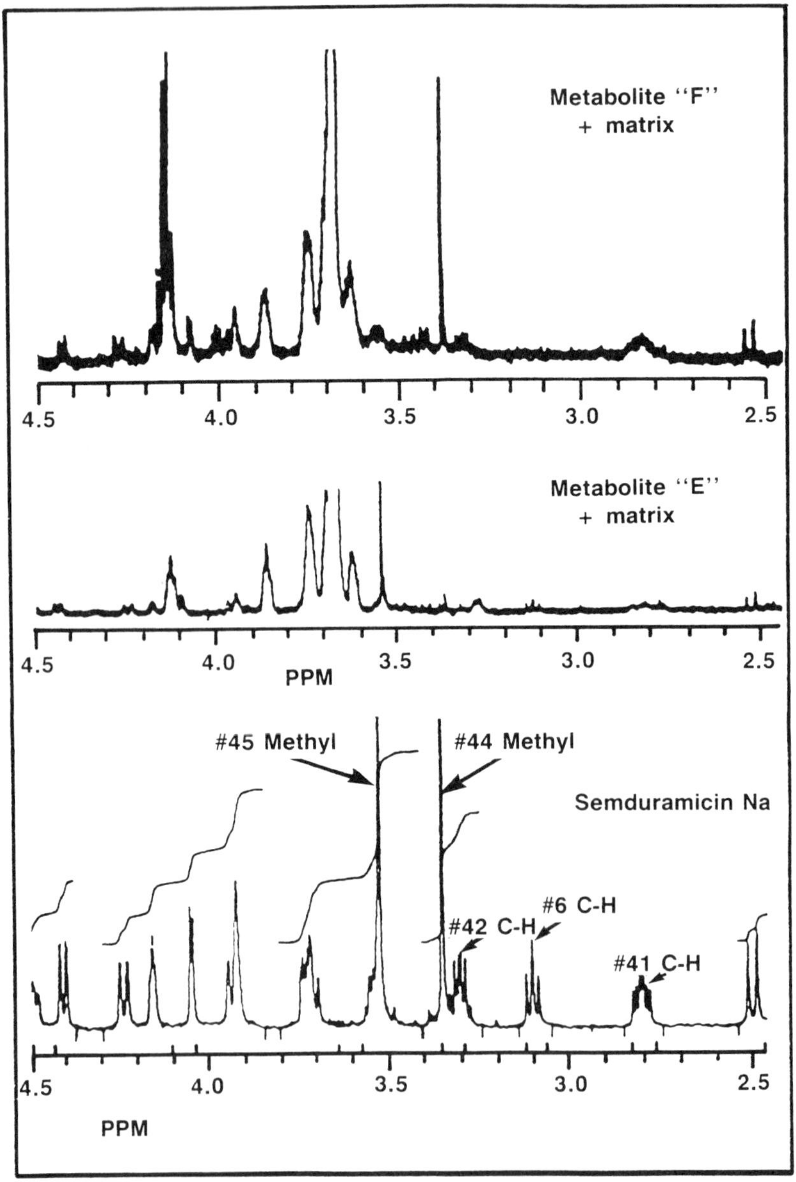

Figure 8. Proton NMR spectra at 500 MHZ of metabolites E, F and semduramicin sodium between 2.5 and 4.5 ppm (shift values in δ ppm vs TMS =0).

distribution and excretion pattern of ^{14}C-semduramicin sodium in the chicken are similar to those of other polyether antibiotics.

As with other polyether antibiotics the major building blocks for the skeletons of all polyethers are acetate, propionate and butyrate (*18*). The biosynthetic incorporation of [2-^{14}C]acetate into semduramicin sodium resulted in the preparation of ^{14}C-semduramicin sodium suitable for radiotracer studies. ^{13}C NMR indicated significant enrichment in many carbons of the polyether rings, but not in methyl, methoxy or sugar substituents. The label was sufficiently distributed to be largely retained in any metabolite containing three or more contiguous carbons from regions other than the sugar substituent itself. Except for the absence of the incorporation of the label into the methyl groups, these results are consistent with the previous report of the incorporation of [2-^{13}C]acetate into maduramicin (*26*).

Following the in-feed administration of ^{14}C-semduramicin sodium at its use level of 25 ppm for 7 consecutive days, the absorption of ^{14}C-semduramicin sodium was characterized by low levels in plasma, muscle, kidney, fat and skin, but by higher levels in bile and liver. Over a 5 day withdrawal period, total residues in each of the edible tissues and bile were depleted to lower levels. The relatively large amount of radioactivity excreted via the bile was consistent with the structure of semduramicin sodium and results reported for monensin (*27*). The drug was eliminated in the excreta and appeared to be at steady state during this dosing period. The consistently low plasma concentrations and the relatively higher residue concentrations in bile and excreta were consistent with metabolic clearance and with an efficient hepatic route of drug elimination.

As defined by reverse isotope dilution analysis and metabolic profiling experiments, unchanged semduramicin sodium was the major component of total residues in liver at the 6 hour withdrawal period. At later withdrawal times, unchanged semduramicin sodium levels declined by a biphasic pattern in liver similar to that of total residues.

Using bile as a source of major metabolites for spectrometric identification, the metabolism of semduramicin is characterized by O-demethylation of the methoxyl groups in the "A" and "G" rings, as previously described for maduramicin (*13-14*) and monensin (*28, 29*).

Because semduramicin has neither strong chromophoric nor electrochemical activity, the formation of a detectable derivative by an in-line, post-column reaction with vanillin was investigated. In combination with a suitable isolation and cleanup procedure, a post column HPLC system, as employed for sodium salinomycin in feeds (*14*), provided a method that was sensitive to 0.010 µg/g in liver. When applied to liver samples taken from birds dosed with ^{14}C-semduramicin, the unchanged drug concentration in liver at the 6, 12 and 24 hour withdrawal periods represented 48, 30 and less than 21 percent of total radioactivity by chemical assay. These findings confirmed that unchanged semduramicin sodium represented a significant portion of residues in liver at early withdrawal times and offered a means for monitoring residues over this period through use of a "marker" (*9-11*).

The utility of determining unchanged semduramicin sodium levels in poultry liver to monitor withdrawal of the drug differentiates this ionophore from others. While relatively high concentrations of total residues have been reported for monensin (*27, 29*), and salinomycin (*30*) in poultry liver, residues of these ionophores are characterized by lower percentages of the unchanged drug. Only in fat, where total residues are low, was a higher percentage of unchanged monensin found. Skin has been designated the target tissue for lasalocid (*30*), and fat for maduramicin (*23, 32*) because these tissues contain higher percentages of the unchanged drug, and, therefore, are useful for monitoring depletion of the drug.

In tandem with a determination of the disposition of ^{14}C-semduramicin sodium in broilers, the depletion of the unchanged drug as a surrogate or marker for total residues was monitored in poultry that received the drug under *ad libitum* administration of feed containing 25 ppm semduramicin for 44 consecutive days. This study confirmed that residues of the drug deplete to levels in the low ng/g range in less than 24 hours. Coupled with results from toxicology studies which establish safe concentrations for residues of semduramicin in edible tissues, a suitable withdrawal period may be established.

Acknowledgments

The authors are grateful to Al Calcagni for preparation and analysis of samples, Justin Stroh for mass spectrometric analyses, and to John Dirlam, Earl Whipple, Jon Schaeffer, T. Schaaf, E. Glazer and R. Ronfeld for thoughtful discussions.

Literature Cited

1. Fernando, M. A.; In *The Biology of Coccidia*; Long, P. L., Ed.; University Park Press: Baltimore, MD, **1982**; pp. 287-327.
2. Reid, W. M.; *Avian Diseases*, **1990**, 34, 509-525.
3. McDougald, L. R.; In *The Biology of Coccidia;* Long, P.L., Ed.; University Park Press, Baltimore, MD, **1982**; pp 373-427.
4. Pressman, B. C.; *Ann. Rev. Biochem.*, **1976**, 45, 501-530.
5. Glazer, E. A.; 530; Cullen, W.P.; 530 Frame, G. M.; Goudie, A. C.; Koss, D. A.; Olson, J. A.; Ricketts, A. P.; Tynan, E. J.; Walshe, N. D.; Wernau, W. C.; Schaaf, T. K.; Annual Meeting of the Society for Industrial Microbiology, July 29-August 3, **1990**, Orlando, FL.
6. *Polyether Antibiotics*, Westley, J. W., Ed.; Marcel Decker, New York, NY, **1982**; Vol. 1.
7. Tynan, E. J.; Nelson, T.H.; Davies, R. A; Wernau, W. C.; Annual Meeting of the Society for Industrial Microbiology, July 29-August 3, **1990**, Orlando, FL.
8. Dorrestein, A. S.; Van Miert, A. S.; *J. Vet. Pharmacol. Therap.*, **1988**, 11, 33-44.
9. Perez, M. K.; *J. Assoc. Off. Anal. Chem.*, **1978**, 61, 1183-1191.
10. Weber, N. E.; *J. Animal Sci.*, **1983**, 56, 244-251.
11. *Criteria and Procedures for Evaluating the Safety of Carcinogenic Residues*; Federal Register (Dec. 31, 1987) 52 FR 49572; 21 CFR Parts 70, 500, 514 and 571.
12. Donoho, A. L.; *J. Animal Sci.*, **1984**, 58, 1528-1539.

13. Brown, M. A.; Rajan, S.; **1986**, *J. Agric. Food Chem.*, 34, 470-472.
14. Stout, S. J.; Wu, J.; daCunha A. R.; King, K. G.; Lee, A.; *J. Agric. Food Chem.*, **1991**, 39, 386-391.
15. Goras, J. T. and Lacourse, W. R.; 706; *J. Assoc. Off. Anal. Chem.*, **1984**, 67, 701-706.
16. Lapointe M. R. and Cohen H.; *J. Assoc. Off. Anal. Chem.*, **1988**, 71, 480-484.
17. Dirlam, J. P.; Cullen, W. P.; Huang, L. H.; Nelson, T. H.; Oscarson, J. R.; Presseau-Linabury, L.; Tynan, E. J.; Whipple, E. B.; *J. Antibiot.* **1991**, 44, 1262-1266.
18. Westley, J. W.; *J. Natural Products*, **1986**, 49, 35.
19. Komarowski, A.; *Chem. Ztg.*, **1903**, 27, 807, 1086.
20. Duke, F. R.; *Anal. Chem.*, **1947**, 19, 661.
21. Blanchflower, W. J.; Desmond, A. R.; and Hamilton, J. T. G.; **1985**, Analyst 110, 1283.
22. Golab, T.; Barton, S. J. and Scroogs, R. E.; *J. Assoc. Off. Anal. Chem.*, **1973**, 56, 171.
23. Siegel, M. M.; McGahren, W. J.; Tomer, K. B.; Chang, T. T.; *Biomedical and Environmental Mass Spectrometry*, **1987**, 14, 29.
24. Stout, S. J.; Wilson, L. A.; Kliener, A. L.; daCunha, A. R.; Franci, T. J.; *Biomed. Mass Spectrom.*, **1989**, 18, 57.
25. Occolowitz, J.L. and Hamill, R. L.; In *Polyether Antibiotics*; Westley, J. W., Ed.; Marcel Decker: New York, NY., **1982**; Vol. 2; pp 197-243.
26. Tsou, H. R.; Rajan S.; Fiala, P. C.; Bullock, M. W.; Borders, D. B.; *J. Antibiot.*, **1984**, 37, 1651.
27. Davidson, K. L.; *J. Agric. Food Chem.* **1984**, 32, 1273.
28. Donoho, A.; Manthey, J.; Occolowitz, J.; Zornes, L.; **1978**, *J. Agric. Food Chem.*, 26, 1090.
29. Donoho, A.L.; Herberg, R. J.; Zornes, L.; Van Duyn, R. L.; **1982**, *J. Agric. Food Chem.*, 30, 909.
30. Dimmena, G. P.; Lyon, F.S.; Thompson, F. M.; Creegan, J. A.; and Wright, G. J.; *J. Agric. Food Chem.*, **1989**, 37, 668.
31. Weiss, G.; and MacDonald, A.; *J. Assoc. Off. Anal. Chem.*, **1985**, 68, 971.
32. *Animal Drugs, Feeds, and Related Products; Maduramicin Ammonium*; Federal Register (Feb. 2,1989) 54 FR 5229; 21 CFR 556.375

RECEIVED February 28, 1992

METHODOLOGY OF XENOBIOTIC METABOLISM

Chapter 6

In Vitro Models for Biotransformation and Toxicity Studies in Farm Animals

Furazolidone and Isolated Pig Hepatocytes

L. A. P. Hoogenboom, M. B. M. Huveneers-Oorsprong, and H. A. Kuiper

State Institute for Quality Control of Agricultural Products (RIKILT–DLO), Bornsesteeg 45, 6708 PD Wageningen, Netherlands

Pig hepatocytes were isolated from liver samples and used for studying the toxicity and biotransformation of veterinary drugs and growth promoting agents, like sulfamethazine, ß-nortestosterone, and furazolidone. In the case of furazolidone, a number of known *in vivo* toxic effects, like the inhibition of pyruvate metabolism and monoamine oxidase activity, could be reproduced and their mechanisms further studied. The biotransformation of furazolidone by these cells resulted in the formation of a number of unknown hydrophilic compounds, in addition to small amounts of the cyano-metabolite and protein-bound metabolites. Using a new analytical method for the detection of these protein-bound metabolites, it was shown that these metabolites should be considered "drug-like". No proof was obtained for the binding of reactive intermediates of furazolidone to the DNA of the cells. It is concluded that *in vitro* models like hepatocytes, prepared from tissues or organs of food-producing animals can be a valuable tool not only for studying the biotransformation of compounds in these animals, but also for investigating the toxic effects and their underlying mechanisms.

The intensification in the breeding of food-producing animals has been accompanied by an increased use of veterinary drugs and growth-promoting agents. However, treatment of animals with drugs may result in residues of both the parent compound and its metabolites in edible products, which may imply a possible health risk for the consumer. Therefore, it is essential to obtain accurate information not only on the toxic properties of the parent drug, but also on its biotransformation products, especially since there are numerous examples where these have been shown to be responsible for the adverse effects of a compound. Since the kind of metabolites are often specific for a certain animal species, such information should preferentially be obtained from studies with target animals. However, experiments with large domestic animals are in general difficult to perform for reasons of costs, duration, housing facilities, and ethics, especially when the use of radiolabeled compounds is required in order to obtain the relevant information. *In vitro* models may offer a good alternative for at least part of the animal studies.

0097–6156/92/0503–0072$06.00/0

In the past decades, there has been a large increase in the development and use of *in vitro* models, not only because of the demand of animal welfare groups to reduce the use of laboratory animals, but also because of a number of scientific and financial advantages. Especially in the case of small laboratory animals, a number of different models have become available for biotransformation and toxicity studies, *e.g.* perfused organs, tissue slices, isolated cells (*e.g.* hepatocytes), cellular fractions (*e.g.* microsomes), and purified enzymes. For a number of reasons (see below) isolated hepatocytes have become one of the most popular *in vitro* models for such studies.

In vitro models derived from material from farm animals have more or less been restricted to tissue homogenates or isolated cell fragments, like microsomes. Only recently the experiences with other animal species have resulted in the isolation and use of hepatocytes and liver slices.

Isolated and Cultured Hepatocytes

In general viable hepatocytes are isolated from livers or liver samples by methods based on the work by Howard *et al.* (*1*), who used an enzymatic treatment of liver slices for the dispersion of the cells. The efficiency of the method has subsequently been improved by the introduction of an *in situ* perfusion, that is commonly used in the case of small laboratory animals (*2*), and the perfusion of the liver with a buffer containing a Ca^{2+}-chelator, like EGTA, prior to enzymatic treatment with collagenase (*3*). More recently, these methods were adapted for the isolation of hepatocytes from liver samples obtained from larger animal species, including man (*4*, *5*, *6*).

Once isolated, hepatocytes can be used either in suspension cultures or as monolayer cultures, where they are allowed to attach to culture dishes. In the case of suspension cultures, a very limited amount of equipment is needed, but a disadvantage is that the cells can only be used for a few hours. Monolayer cultures stay viable for several days or weeks, but their use requires the presence of tissue culture facilities, in order to work under sterile conditions.

The applicability of hepatocytes for various uses has been extensively reviewed (*7*, *8*, *9*, *10*). Major advantages and disadvantages of the use of this model are described below.

Full Biotransformation Potential. In comparison with cellular fractions, freshly isolated hepatocytes contain a complete set of biotransformation enzymes in the original cellular configuration. As a result, compounds may be transformed by a number of subsequent metabolic steps, very similar to those *in vivo*. In addition, the use of expensive cofactors is superfluous, due to the intact metabolic functioning of the cells. However, as a consequence, cellular fractions are sometimes more suitable for studying a specific biotransformation step or for the detection of reactive intermediates.

Intact Cellular Structure. The intact cellular structure offers the opportunity to use hepatocytes for studying (toxic) effects of compounds on the functioning and viability of the cells, possibly in relation to the biotransformation of another compound (interaction studies). Although in theory each cellular function can be used to determine the toxicity of a compound, a number of tests have gained a more widespread popularity.

Many Experimental Units. An important advantage of hepatocytes both over the intact animal, and perfused organ, is that cells isolated from one liver or liver sample can be divided over a large number of experimental units. This is very useful for studying *e.g.* dose- and time- relationships in the biotransformation or toxic effects of a compound, or for comparing the effects of different compounds, and studying the interactions between a number of compounds. The fact that the system is easy to handle and small in volume, makes it very suitable for the use of radiolabeled or rare and expensive compounds.

Dedifferentiation. A major problem encountered with monolayer cultures of hepatocytes, is that their prolonged use is limited by a change in the activity of certain biotransformation enzymes during the ageing of the cells (for review see (*11*)). Initially, this was shown for the very important group of cytochrome P-450 related enzyme activities, which often decrease to 50% or more within 24 hr. Subsequent studies also reported changes, both increases and decreases, for other enzymes and enzymatic activities (*12*, *13*, *14*, *15*). The problem is not restricted to hepatocytes from rats, although in the case of human hepatocytes, the decrease in cytochrome P-450 levels appears to be much slower (*6*, *16*).

A number of possible solutions have been proposed to solve the problem, like the use of media supplemented with certain hormones (*17*), nicotinamide, or metyrapone, or media free of cystine and cysteine, but supplemented with aminolaevulinic acid (*11*). However, the application of such media has been a limited success since the apparently stable levels of cytochrome P-450 were accompanied by a selective change in the pattern of the different isoenzymes. A more promising approach appears to be the long-term cocultivation of hepatocytes with a rat epithelial cell-line, resulting in steady state levels of cytochrome P-450 after an initial decline (*6*, *18*). However, the results of a recent study by Rogiers *et al.* (*19*), indicate that this technique does not offer a complete solution for the selective shift in the cytochrome P-450 isoenzyme patterns.

In general, one should be aware of the possible decrease in certain enzyme activities, when using older cultures. However, it should be stressed that, although a decline in enzyme activities might have consequences for the rate of conversion of a compound, it does not necessarily result in a change in metabolite patterns.

Selective Biotransformation Potential. A number of factors might limit the possible extrapolation of results to the *in vivo* situation. Firstly, compounds may be degraded by different cells or mechanisms, *e.g.* due to acid conditions in the stomach, or by bacterial degradation in the lower parts of the G.I.-tract. Also, other organs may be involved in the biotransformation of a compound, possibly resulting in very specific metabolites. On the other hand, the isolation of hepatocytes results both in the disruption of the typical liver structure, and a selection of cell-types. This may be an important advantage of liver slices, especially now that their preparation has become much easier, faster and more reproducible (*20*, *21*, *22*, *23*, *24*). Last but not least, monolayer cultures like most other *in vitro* models, are not a dynamic system, allowing for example the degradation of metabolites, that might otherwise be excreted before further degradation.

The Use of *in vitro* Models to Study Species-differences

Both quantitative and qualitative differences have been reported in the biotransformation of xenobiotics by different animal species (*25*, *26*). Initially, indications for this phenomenon came primarily from toxicity studies with small laboratory animals, showing species-differences in the sensitivity towards the toxic effects of a compound. *In vivo* studies have for example revealed a large difference between rats and mice in their sensitivity to the hepatotoxic and carcinogenic properties of aflatoxin B_1 (*27*). More recently, species-related differences in the biotransformation of compounds were shown in studies with isolated cellular fractions from livers of different animal species, including farm animals (*28*, *29*, *30*, *31*, *32*).

The development of procedures for the isolation of hepatocytes from livers of different species, including man, have allowed the use of this model for comparative studies on the species-related biotransformation and toxicity of xenobiotics. *In vitro* studies with aflatoxin B_1 showed that rat hepatocytes were more sensitive than those from mice, as became apparent by the death of the cells at much lower concentrations, the much higher levels of binding of metabolites of aflatoxin B1 to TCA-precipitable cell-material (*33*, *34*), and the much higher levels of unscheduled DNA-synthesis (*35*). In addition, the mycotoxin caused segregation of nucleolar components in rat hepatocytes, but not in cells from mice (*36*).

Several research groups have demonstrated that species differences in the biotransformation of drugs with diverse but known *in vivo* metabolite patterns, could be reproduced with hepatocytes. Hepatocytes from diverse species like rats, rabbits, dogs, monkeys and man have been used to study the biotransformation of tolbutamide (*37*), amphetamine (*38*), ketitofen (*39*) and diazepam (*40*). These and other comparative studies again demonstrate that in general extrapolation of data obtained with cellular fractions, hepatocytes or intact animals from one species to another species is not possible, and stresses the need for the use of target-animals for risk-assessment purposes.

Isolation and Use of Hepatocytes from Large Domestic Animals

The possible isolation of hepatocytes from liver samples of large domestic animals has been the subject of a limited number of studies. Suspension cultures of hepatocytes, isolated from liver samples of sheep (*41*, *42*, *43*, *44*), goats (*45*, *46*, *47*), and cows (*46*), were used to study the gluconeogenesis from propionate and butyrate, a process that is relatively specific for ruminants. Monolayer cultures of pig hepatocytes, isolated by an *in situ* perfusion from livers of neonatal animals, were used to study the catabolism of lipoproteins (*48*, *49*), regarding the good similarity between humans and pigs with respect to the importance of the LDL particle. In the same field of interest, monolayer cultures of pig hepatocytes, isolated from liver samples of young animals (up to 7 weeks), were used to investigate the feedback regulation of bile acid synthesis, being an important factor in the elimination of cholesterol from the body (*50*, *51*). Finally, suspension cultures of hepatocytes, isolated by an *in situ* perfusion from 2-15 days old piglets were used to study the effects of fasting on the gluconeogenesis (*52*).

Investigators have only recently become aware of the opportunities to use hepatocytes from large domestic animals for studying the possible species-specific biotransformation and toxicity of compounds used in veterinary practice. Suolinna and Winberg (*53*) studied the deethylation of 7-ethoxycoumarin and the conjugation of methyl-umbelliferone by suspension cultures of bovine hepatocytes isolated from liver samples obtained at the slaughterhouse from fetal or adult (1.5-2 years) animals. Shull *et al.* (*54*, *55*), introduced the surgical removal of the caudate lobe of the bovine liver, and its subsequent use for the isolation of hepatocytes. Subsequently the biotransformation of 7-ethoxycoumarin and the pesticide aldrin was studied in monolayer cultures. Mennes *et al.* (*16*) included sheep hepatocytes in a study with rat, hamster, dog, monkey and human hepatocytes, in order to compare the decline of cytochrome P-450 levels in monolayer cultures. Recently Pool *et al.* (*56*) used suspension cultures of pig hepatocytes obtained from 12 months old animals, to study the effect of a number of nitrosamines on the frequency of single strand breaks in the DNA.

These studies clearly demonstrate the possibility to isolate viable hepatocytes from large domestic animals, but the use of these cells for biotransformation and toxicity studies of compounds used in veterinary practice, has not been thoroughly studied. We have investigated, therefore, in our laboratory the possibility of isolating viable hepatocytes from pig liver samples, obtained at the slaughterhouse and the subsequent use of these cells for biotransformation and toxicity studies. In particular we have concentrated on the formation of metabolites of veterinary drugs that might be present as residues in edible products.

Isolation and Culturing of Pig Hepatocytes. Hepatocytes were isolated from liver samples of pigs (Yorkshire x Dutch Landrace) aged 4-15 months, by a three step perfusion based on the method of Seglen (*3*), as described previously (*57*). Animals were killed by electrocution followed by exsanguination. A distal part of a liver lobe was obtained (about 150 g), and washed with ice-cold NaCl (0.9%). The sample was transported to the laboratory in ice-cold NaCl, and connected to an oxygenator (*58*). The liver sample was perfused with 1 litre of EGTA-buffer, 1 litre of the same buffer without EGTA, and finally 0.4 l of a buffer containing 0.05% of collagenase (type I, Sigma) for about 30 min under recirculation. During this period the liver swelled and the connective tissue started to leak. Subsequently the sample was quickly cut into small parts and the suspension filtered over a Büchner funnel, to remove the larger parts, and then a 250 μm nylon filter. The cells were collected by centrifugation, washed (3x), and filtered through a 250 μm nylon filter. Cell suspensions were diluted to a density of 1.25-1.5 million viable cells per ml of Williams' medium E, supplemented with 5% FCS, 0.5 μg insulin per ml, 50 IU penicillin per ml and 50 μg streptomycin per ml, seeded in culture dishes and incubated in an incubator at 38 °C, 5% CO_2 and high humidity. After 4 hr the medium was replaced by new medium.

In most cases the method resulted in a yield of 2.5-5 million viable cells per gram of liver, with a viability of 90-95% (trypan-blue excluding cells). Cells normally readily attached to uncoated culture dishes, and formed monolayer cultures within 24 hr. However, in some cases less viable cells or cells with poor plating efficiency were obtained, without an obvious reason.

Biotransformation of Sulfamethazine and ß-Nortestosterone by Pig Hepatocytes. Incubation of pig hepatocytes, isolated from livers of a number of different animals, with the antibacterial agent sulfamethazine resulted in the formation of a single metabolite, N^4-acetyl-sulfamethazine (*57*), as previously reported *in vivo* (*59*). In addition it was shown that these cells were capable of deacetylating part of the N^4-acetyl-metabolite to the parent compound.

The anabolic steroid ß-nortestosterone was rapidly metabolized by the cells, initially resulting in the formation of norandrostenedione, which was further transformed into the glucuronide of 15α-hydroxy-norandrostenedione (*60*). The same metabolites were subsequently shown to be present in urine samples of pigs treated with ß-nortestosterone.

Toxicity and Biotransformation Studies with Furazolidone. Most studies with pig hepatocytes were performed with the nitrofuran drug furazolidone. Previous studies in our laboratory with ^{14}C-labeled furazolidone, revealed that this drug was extensively biotransformed in piglets, partly resulting in the presence of unextractable radiolabeled compounds in the protein fraction (*61*). Studies with pig liver microsomes (*62*) showed that furazolidone is most likely metabolized by reduction of the nitro group, initially resulting in the formation of a radical nitro-anion (Figure 1), and subsequently an acrylonitrile-derivative. The latter is thought to be responsible for the binding with thiol-group containing compounds, like glutathione, cysteine or protein, and possibly DNA, but might also be further reduced to the known *in vitro* and *in vivo* cyano-metabolite of furazolidone.

Regarding the persistence of bound residues, as compared to the parent compound and other metabolites, and their high bioavailibility when fed to rats (*63*), special attention was paid to their identity and possible toxic properties. In particular the possible presence of reversible protein-bound metabolites in tissues of treated animals was hypothesized to imply a possible health-risk for the consumer (*63*). The issue of bound-residues of furazolidone and other drugs has long been one of the most controversial and difficult problems in the field of residue toxicology (*64*, *65*).

Toxicity Studies. Initially a number of more commonly used parameters, like the leakage of the cytosolic enzyme lactate dehydrogenase (LDH) out of the cells, the *de novo* synthesis of proteins, and the intracellular levels of reduced and oxidized glutathione were used to study the cytotoxicity of furazolidone (*66*). As shown in Table I, there was a slight increase in the LDH-activity in the medium at the highest drug concentration, but only when corrected for the total activity on the plate. Therefore, this apparent increase does not appear to be due to membrane damage, but more likely to a decreased synthesis of the enzyme, especially considering the decreased protein-synthesis observed at higher drug concentrations (Table I). A decreased protein-synthesis was especially apparent for proteins excreted into the medium. It is unclear whether there is a relation between this effect and the decreased plasma protein levels that have been measured in the case of chickens (*67*), turkeys (*68*, *69*), and goats (*70*) treated with furazolidone.

In agreement with the role of GSH in the deactivation of reactive oxygen-species formed by redox-cycling of the initially formed radical nitro-anion (Figure 1) (*71*), there was a small increase in intracellular GSSG-levels (*66*). Unexpectedly, this was not accompanied by a decrease in the GSH-levels. However, subsequent experiments revealed that exposure to furazolidone does result in an increased loss of GSH by the cells, but that the loss is compensated by an increased GSH-synthesis (Hoogenboom, L.A.P.; Kammen, M. van; Huveneers-

Table I. Effect of furazolidone on LDH-leakage and protein synthesis in pig hepatocytes (mean ± SEM; n=3). Adapted from ref. 66

[Fur]	LDH (absolute)		LDH	Protein synthesis	
	medium	cells	medium	medium	cells
(μM)	(U/mg cellular protein) (x 0.001)		(% of total)	(% of total)	
0	29 ± 3	0.62 ± 0.02	4.45 ± 0.24	100 ± 3	100 ± 4
15	28 ± 1	0.67 ± 0.02	4.04 ± 0.04	111 ± 4	114 ± 1
50	29 ± 1	0.72 ± 0.01	3.91 ± 0.11	85 ± 4	114 ± 2
150	28 ± 0	0.70 ± 0.01	3.68 ± 0.14	58 ± 4	100 ± 4
500	31 ± 1	0.48 ± 0.02	6.03 ± 0.11	42 ± 5	59 ± 1

Oorsprong, M.B.M.; Kuiper, H.A. *Toxicology In Vitro*, in press). This could be demonstrated by using buthionine-S-sulfoximine, a specific inhibitor of GSH-synthesis, but also by using cultures containing ^{35}S-labeled GSH, achieved by preincubation with ^{35}S-methionine. In the latter case, there was a dose-related, increased loss of radiolabeled GSH, again in the absence of an effect on absolute GSH-levels (Figure 2). This effect may offer an explanation for the increased GSH-levels observed in livers from salmons and chars treated with furazolidone (*72*).

Two other, more specific parameters were used to study the toxicity of furazolidone. As shown in Figure 3, the drug had a marked inhibitory effect on the metabolism of pyruvate, resulting in the accumulation of pyruvate and lactate in the medium. When the medium containing the drug was replaced by control medium, the cells gradually, but slowly recovered from the effect. The effect was also observed with the related nitrofuran drugs, nitrofurantoin, furaltadone and nitrofurazone, and turned out to be the most sensitive parameter used. Previously, Paul *et al.* (*73*) observed a very similar, slowly recovering effect in tissues of rats treated with nitrofuran drugs, and hypothesized that the effect is caused by an irreversible binding of the parent drug or a metabolite to one of the enzymes in the pyruvate dehydrogenase complex (*74*). The effect might be responsible for the polyneuritis, which is often observed in animals and humans, treated with nitrofuran drugs.

A rather specific effect observed in tissues from animals treated with furazolidone, but not other nitrofurans, is the irreversible inhibition of monoamine oxidase (MAO) activity (*75*, *76*, *77*, *78*). There is good evidence that the inhibition is due to a metabolite formed from the 3-amino-2-oxazolidinone side-chain of the drug (*76*). However, the site of formation of the final metabolite remains a point of discussion, especially since the effect was not observed *in vitro* with isolated mitochondria in the absence or presence of a liver homogenate. Using hepatocytes, we have been able to demonstrate the existence of a second, reversible type of MAO-inhibition by furazolidone, but also furaltadone and nitrofurazone (*79*). However, an irreversible type of MAO-inhibition was only observed after incubation of cells with the AOZ side-chain of furazolidone, and not with the side-chain of furaltadone. From these data it became clear that once the side-chain is released by *e.g.* acid hydrolysis, it can be metabolized into an active MAO-inhibiting compound by mammalian cells.

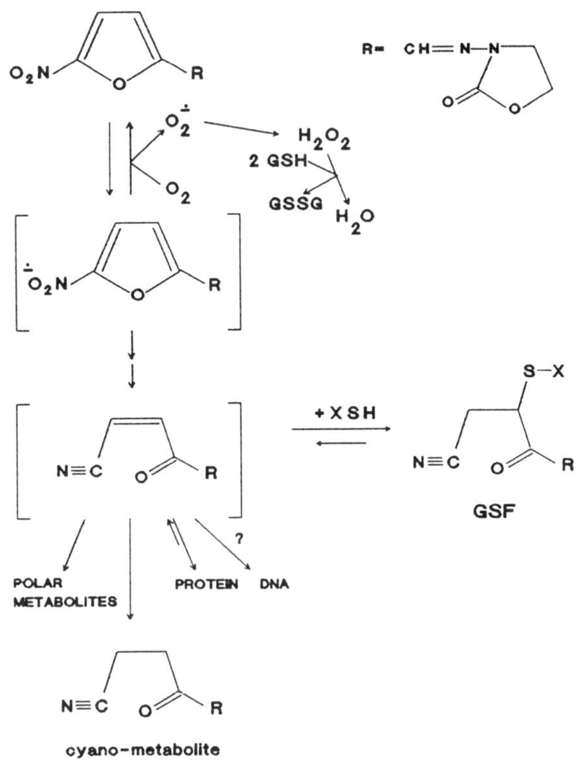

Figure 1. Proposed biotransformation pathway of furazolidone by pig liver microsomes (*62*).

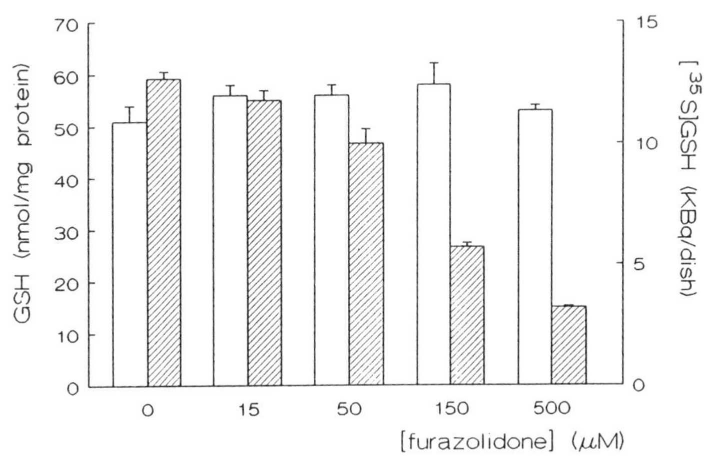

Figure 2. Effect of furazolidone on intracellular GSH-levels. Prior to the exposure, cells were incubated with ^{35}S-methionine, resulting in the labelling of GSH. Following exposure, amounts of both total (▭) and radiolabeled (▨) GSH were determined.

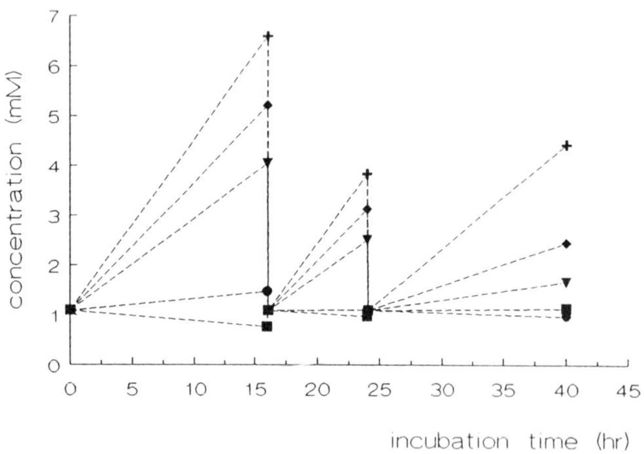

Figure 3. Accumulation of pyruvate and lactate in the medium following exposure of cells to 0 ■ , 15 ● , 50 ▼ , 150 ◆ , or 500 + μM furazolidone. Exposure was started at t=0 and ended at t=16 hr. At t=16 and t=24 hr the medium was replaced with control medium. Adapted from ref. 66.

Biotransformation Studies. Incubation of 24-hr old monolayer cultures with furazolidone resulted in degradation of the parent compound into a large number of more hydrophilic metabolites (Figure 4A). Only the cyano-metabolite (CN), which was previously detected as a minor metabolite in the blood of treated pigs (*80*), could be identified. Subsequent studies with this metabolite showed its further degradation in the cells into a number of other metabolites (Figure 4B); it was hitherto assumed to be a stable terminal metabolite.

No evidence could be obtained for the formation of a reversible glutathione or other thiol-conjugate, as previously shown in studies with pig liver microsomes (Figure 1). A possible explanation might be the reactivity of these type of metabolites, possibly resulting in their binding to cellular proteins. However, if so, this did not result in the formation of reversible protein-thiol conjugates, as shown by the absence of a mercaptoethanol-conjugate after incubation of cellular protein with mercaptoethanol. From these studies, it was concluded that reversible thiol-conjugates appear to be of no relevance with respect to the consumer.

The biotransformation of furazolidone resulted in the formation of so-called protein-bound metabolites as found *in vivo* (*61*), judged by the presence of unextractable radiolabeled compounds in the protein fraction (Figure 5). The formation of these compounds was both dose- and time-related. After replacement of the medium containing the drug with control medium, there was a gradual decrease in the levels of bound metabolites, accompanied by the excretion of hydrophilic compounds into the medium. There was no clear difference in the formation and degradation pattern of bound residues when two different ^{14}C-furazolidone preparations, labeled in either the nitrofuran-ring or the 3-amino-2-oxazolidinone ring, were used. These results indicate that the formation of bound metabolites is due to the binding of reactive metabolites to proteins (most likely via the nitrofuran-ring), and is not the result of the incorporation of small labeled compounds into amino-acids and subsequently proteins. In order to prove that these bound residues should be considered "drug-like", a method was developed for the release and detection of the AOZ side-chain from protein-bound metabolites (*81*). The method was based on the acid hydrolysis of the azomethine bond (R_1-C=N-R_2), followed by the trapping and derivatization of AOZ with 2-nitro-benzaldehyde (NBA) to give 3[(2-nitrophenyl)methylene]-amino]-2-oxazolidinone (NPAOZ). Treatment of protein from cells, that had been incubated with furazolidone labeled in the AOZ side-chain, with 0.1 N HCl and 0.5 mM NBA, resulted in the formation of NPAOZ which could easily be extracted with ethylacetate and analyzed by HPLC, using either UV or radioactivity detection. Furthermore, the compound had a UV-spectrum identical to that of authentic NPAOZ. Subsequent studies revealed that under optimal conditions, AOZ could be released from 80% of the bound metabolites. The method was also successfully applied to liver samples from piglets treated with ^{14}C-furazolidone (*61*), although in this case AOZ could be released from only 25% of the bound metabolites. It was hypothesized that the discrepancy may be related to the fact that *in vivo* only part of the bound metabolites is due to the formation of reactive metabolites by reduction of the nitro-group (Figure 1). The rest might be due to either endogenous incorporation of labelled fragments, or to the formation of reactive metabolites from the free AOZ side-chain, possibly via a hydrazine, being the presumed mechanism behind the MAO-inhibition (*77*). In the case of hepatocytes, only the first class of metabolites would be expected, since the cells appeared to be unable to hydrolyse the azomethine bond (see above).

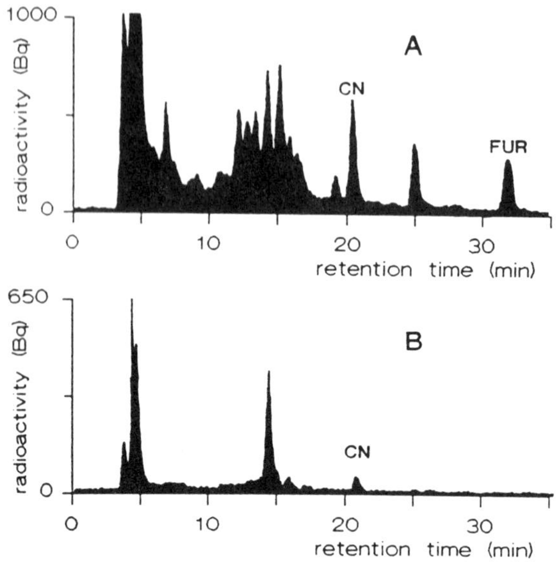

Figure 4. Biotransformation of furazolidone (A) and its cyano-metabolite (B) by pig hepatocytes. Media samples were analyzed after incubation of the cells with 50 μM [^{14}C]furazolidone for 26 hr, or 5 μM of [^{14}C]cyano-metabolite for 36 hr.

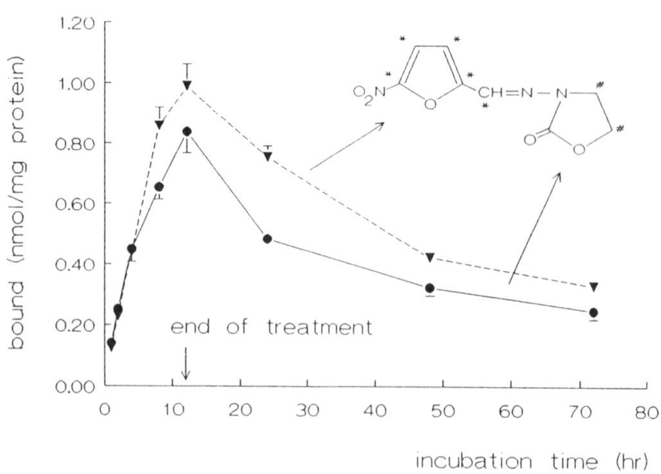

Figure 5. Formation of protein-bound metabolites in hepatocytes exposed to 50 μM furazolidone, radiolabeled in the nitrofuran (▼) or AOZ part (●) of the molecule as indicated. After t=12 hr, exposure was ended by replacing the drug containing medium by control medium. Adapted from ref. 81.

DNA-binding Studies. Another important subject in the field of residue toxicology, is the possible binding of drugs or their metabolites to DNA, making it a possible genotoxic agent. In the past furazolidone has been shown to induce tumours in rats and mice (*82*), and in addition gives clear positive results in the *Salmonella* mutagenicity test (*83*). Furthermore, Vroomen has demonstrated the presence of radiolabel in DNA purified from tissues of piglets treated with ^{14}C-furazolidone (*61*). When DNA was isolated from pig hepatocytes that had been incubated with ^{14}C-furazolidone, small amounts of radiolabel could be detected in this fraction. A major problem was, however, to exclude the possible contamination of this fraction with protein, considering the previously detected protein-binding. Therefore, a second experiment was performed in which part of the cells were incubated with ^{14}C-furazolidone, and part of the cells with ^{14}C-leucine, an essential amino-acid, exclusively used to synthesize proteins. As shown in Table II, radiolabeled substances could be detected in the purified DNA-fraction, both in the case of cells incubated with the drug and in those incubated with the aminoacid. From the leucine data it can be calculated that the DNA-fraction still contains about 6 μg of the original 37 mg of cellular protein, and therefore it can be concluded that in the case of furazolidone, all radiolabeled material in the DNA is most likely derived from this protein contamination. These results are in agreement with previous results from studies with microsomes incubated with the drug in the presence of calf thymus DNA, also failing to detect DNA-bound metabolites (*84*). The discrepancy between the *in vitro* and *in vivo* data might be explained by a possible protein-contamination in the latter case, despite a higher purity as indicated by a A260/A280 ratio of 1.86. A second explanation may be, that the DNA-binding *in vivo* is caused by reactive intermediates, derived from the 3-amino-2-oxazolidinone part of the molecule.

Table II DNA-binding studies on furazolidone. Pig hepatocytes were incubated with 50 μM ^{14}C-furazolidone for 16 hr (starting 26 hr after isolation), or ^{14}C-leucine for 40 hr (starting 4 hr after isolation). Cells were homogenized, the chromatin collected by centrifugation and DNA purified by phenolic extraction

	Protein-binding (dpm/μg protein)[1]		DNA-binding	
	cell-homogenate	chromatin	μg DNA[2]	dpm/μg DNA
furazolidone (n=3)	100.6	72.4	28.0	12.0
leucine (n=2)	54.4	44.2	21.7	12.6

[1] assuming that all label is bound to protein, and not DNA
[2] A260/A280 was 1.67

Conclusions

The results of these studies show that isolated hepatocytes are a valuable tool for studying the biotransformation and toxicity of xenobiotics. Especially in the case of larger animals, such models may be used to obtain information, that is otherwise hardly accessible. Firstly, metabolites can be readily isolated for characterization and identification, and subsequently use for comparison in *in vivo* studies. Furthermore, as shown in the case of ß-nortestosterone and furazolidone, our studies also resulted in the development of new analytical methods, based on the detection of these metabolites instead of parent compounds.

An even greater advantage is the opportunity to use the model for studying the biological properties of drugs and their metabolites. This is important with respect to the consumer, regarding the possible presence of residues in edible tissues, but such studies may also help to obtain information on the mechanisms behind certain adverse effects in target animals, and be of use in the selection of old drugs or the development of new drugs. In particular the possibility of manipulating certain culture conditions, offers great opportunities. As a result we were able to show an increased GSH-turnover after exposure of cells to furazolidone (Figure 2), the reversibility or irreversibility of certain toxic effects (Figure 3), the biotransformation of the cyano-metabolite, which was formerly thought to be a terminal metabolite (Figure 4B), and the fact that the presence of radiolabeled compounds in the purified DNA-fraction was most likely due to protein-contamination (Table II).

Our studies also show that special care should be paid with respect to the extrapolation of data to the *in vivo* situation and validation of results in studies with animals will currently remain necessary. However, further experience with these and other complementary *in vitro* models might eventually result in a significant reduction in the number of animals needed for these experiments.

Literature Cited

1 Howard, R.B.; Christensen, A.K.; Gibbs, F.A.; Pesch, L.A. *J. Cell Biol.* **1967**, *35*, 675-684.

2 Berry, M.N.; Friend, D.S. *J. Cell Biol.* **1969**, *43*, 506-520.

3 Seglen, P.O. In *Methods in cell biology*; Prescot, D.M., Ed.; Academic Press: New York, **1976**, Vol. 13, pp. 29-83.

4 Reese, J.A.; Byard, J.L. *In Vitro* **1981**, *17*, 935-940.

5 Strom, S.C.; Jirtle, R.L.; Jones, R.S.; Novicki, D.L.; Rosenberg, M.R.; Novotny, A.; Irons, G.; McLain, J.R.; Michalopoulos, G. *J. Natl Cancer Inst.* **1982**, *68*, 771-778.

6 Guillouzo, A.; Bégué, J.-M.; Campion, J.-P.; Gascoin, M.-N.; Guguen-Guillouzo, C. *Xenobiotica* **1985**, *15*, 635-641.

7 Fry, J.R.; Bridges, J.W. *Rev. Biochem. Toxic.* **1979**, *1*, 201-247.

8 Moldéus, P.; Jernstrom, B.; Dawson, J. *Rev. Biochem. Toxicol.* **1983**, *5*, 239-265.

9 Guillouzo, A. In *Isolated and cultured hepatocytes*; Guillouzo, A.; Guguen-Guillouzo, C., Eds. John Libbey, Paris, **1986**, pp 313-332.

10 Chenery, R.J. In *Progress in Drug Metabolism*; Gibson, G., Ed.; Taylor & Francis: London, **1988**, Vol. 11; pp 217-265.

11 Paine, A.J. *Chem.-Biol. Interactions* **1990**, *74*, 1-31.

12 Grant, M.H.; Melvin, M.A.L.; Shaw, P.; Melvin, W.T.; Burke, M.D. *FEBS letters* **1985**, *190*, 99-103.

13 Croci, T.; Williams, G.M. *Biochem. Pharmacol.* **1985**, *34*, 3029-3035.
14 Grant, M.H.; Hawksworth, G.M. *Biochem. Pharmacol.* **1986**, *35*, 2979-2982.
15 Vandenberghe, Y.; Glaise, D.; Guillouzo, A.; Ketterer, B. *Biochem. Pharmacol.* **1988**, *37*, 2482-2485.
16 Mennes, W.C.; Blaauboer, B.J.; Holsteijn, C.W.M. van, In *Liver Cells and Drugs*; Guillouzo, A. Ed.; Colloque INSERM; John Libbey: London, **1988**, Vol. 164; pp 351-355.
17 Decad, G.M.; Hsieh, D.P.H.; Byard, J.L. *Biochem. Biophys. Res. Commun.* **1977**, *78*, 279-287.
18 Bégué, J.-M.; Guguen-Guillouzo, C.; Pasdeloup, N.; Guillouzo, A. *Hepatology* **1984**, *4*, 839-842.
19 Rogiers, V.; Vandenberghe, Y; Callaerts, A.; Verleye, G.; Cornet, M.; Mertens, K.; Sonck, W.; Vercruysse, A. *Biochem. Pharmacol.* **1990**, *40*, 1701-1706.
20 Wormser, U.; Ben Zakine, S.; Stivelband, E.; Eizen, O.; Nyska, A. *Toxic. in Vitro* **1990**, *4*, 783-789.
21 Barr, J.; Weir, A.J.; Brendel, K.; Sipes, I.G. *Xenobiotica* **1991**, *21*, 331-339.
22 Barr, J.; Weir, A.J.; Brendel, K.; Sipes, I.G. *Xenobiotica* **1991**, *21*, 341-350.
23 Fisher, R.; Naut, H.; Gandolfi, A.J.; Brendel, K. *Toxic. In Vitro* **1991**, *5*, 201-205.
24 Gunawardhana, L.; Barr, J.; Weir, A.J.; Brendel, K.; Sipes, I.G. *Drug Metab. Dispos.* **1991**, *19*, 648-654.
25 Williams, R.T. *Biochem. Soc. Trans.* **1974**, *2*, 359-377.
26 Kato, R. *Pharmac. Ther.* **1979**, *6*, 41-98.
27 Croy, R.G.; Essigman, J.M.; Wogan, G.N. *Basic Life Sci.* **1983**, *24*, 49-62.
28 Patterson, D.S.P.; Roberts, B.A. *Res. Vet. Sci.* **1970**, *11*, 399-401.
29 Souhaili-El-Amri, H.; Batt, A.M.; Siest, G. *Xenobiotica* **1986**, *16*, 351-358.
30 Watkins, J.B.; Klaassen, C.D. *J. Animal Sci.* **1986**, *63*, 933-942.
31 Dalvi, R.R.; Nunn, V.A.; Juskevich, J. *J. Vet. Pharmacol. Therap.* **1987**, *10*, 164-168.
32 Short, C.R.; Flory, W.; Hsieh, L.C.; Aranas, T.; Ou, S.P.; Weissinger, J. *Comp. Biochem. Physiol.* **1988**, *91C*, 419-424.
33 Decad, G.M.; Dougherty, K.K.; Hsieh, D.P.H.; Byard, J.L. *Toxicol. Appl. Pharmacol.* **1979**, *50*, 429-436.
34 Hanigan, H.M.; Laishes, B.A. *Toxicology* **1984**, *30*, 185-193.
35 Steinmetz, K.L.; Green, C.E.; Bakke, J.P.; Spak, D.K.; Mirsalis, J.C. *Mutation Res.* **1988**, *206*, 91-102.
36 Cole, K.E.; Hsu, I.-C.; Trump, B.F. *Cancer Res.* **1986**, *46*, 1290-1296.
37 Gee, S.J.; Green, C.E.; Tyson, C.A. *Drug Metab. Dispos.* **1984**, *12*, 174-178.
38 Green, C.E.; LeValley, S.E.; Tyson, C.A. *J. Pharmacol. Exp. Ther.* **1986**, *237*, 931-936.
39 Le Bigot, J.F.; Begue, J.M.; Kiechel, J.R.; Guillouzo, A. *Life Sciences* **1987**, *40*, 883-890.
40 Seddon, T.; Michelle, I.; Chenery, R.J. *Biochem. Pharmacol.* **1989**, *38*, 1657--1665.
41 Clark, M.G.; Fisell, O.H.; Jarret, I.G. *Biochem. J.* **1976**, *156*, 671-680.
42 Démigne, C.; Yacoub, C.; Remesy, C.; Fafournoux, P. *Biochim. Biophys. Acta* **1986**, *875*, 535-542.

43 Chow, J.C.; Planck-Meyer, C.; Jesse, B.W. *J. Dairy Sci.* **1990**, *73*, 683-689.
44 Upreti, G.C.; Riches, P.C.; Ratcliff, R.A. *Small Ruminants Research* **1991**, *4*, 175-187.
45 Aiello, R.J.; Armentano, L.E. Comp. *Biochem. Physiol.* **1987**, *88B*, 193-201.
46 Aiello, R.J.; Armentano, L.E. *J. Dairy Sci.* **1987**, *70*, 2504-2510.
47 Aiello, R.J.; Armentano, L.E.; Bertics, S.J.; Murphy, A.T. *J. Dairy Sci.* **1989**, *72*, 942-949.
48 Pangburn, S.H.; Newton, R.S.; Chang, C-M.; Weinstein, D.B.; Steinberg, D. *J. Biol. Chem.* **1981**, *256*, 3340-3347.
49 Bachorik, P.S.; Franklin, F.A.; Virgil, D.G.; Kwiterovich, P.O. *Biochemistry* **1982**, *21*, 5675-5684.
50 Kwekkeboom, J.; Voorthuizen, E.M. van; Princen, H.M.G.; Kempen, H.J.M. *Biochem. Biophys. Res. Commun.* **1988**, *155*, 850-856.
51 Kwekkeboom, J.; Princen, H.M.G.; Voorthuizen, E.M. van; Meijer, P.; Kempen, J.M. *Biochem. Biophys. Res. Commun.* **1989**, *162*, 619-625.
52 Pegorier, J.P.; Duee, P.H.; Girard, J.; Peret, J. *J. Nutr.* **1982**, *112*, 1038-1046.
53 Suolinna, E.-M.; Winberg, A.-M. *Drug Metab. Dispos.* **1985**, *13*, 722-724.
54 Shull, L.R.; Kirsch, D.G.; Lohse, C.K.; Carlson, G.P.; Doody, L.K.; Wisniewski, J.A. *Am. J. Vet. Res.* **1986**, *47*, 2043-2052.
55 Shull, L.R.; Kirsch, D.G.; Lohse, C.L.; Wisniewski, J.A. *Xenobiotica* **1987**, *17*, 345-363.
56 Pool, B.L.; Brendler, S.Y.; Liegibel, U.M.; Tompa, A.; Schmezer, P. *Environm. Molecular Mutag.* **1990**, *5*, 24-35.
57 Hoogenboom, L.A.P.; Pastoor, F.J.H.; Clous, W.E.; Hesse, S.E.; Kuiper, H.A. *Xenobiotica* **1989**, *19*, 1207-1219.
58 Moldéus, P.; Hogberg, J.; Orrenius, S. *Meth. Enzymol.* **1978**, *52*, 60-71.
59 Paulson, G.D.; Giddings, J.M.; Lamoureux, C.H.; Mansager, E.R.; Struble, C.B. *Drug Metab. Dispos.* **1981**, *9*, 142-146.
60 Hoogenboom, L.A.P.; Berghmans, M.C.J.; Veldhuizen, A. van;; Kuiper, H.A. *Drug Metab. Dispos.* **1990**, *18*, 999-1004.
61 Vroomen, L.H.M.; Berghmans, M.C.J.; Leeuwen, P. van; Struijs, T.D.B. van der; Vries, P.H.U. de; Kuiper, H.A. *Fd Add. Contam.* **1986**, *3*, 331-346.
62 Vroomen, L.H.M.; Berghmans, M.C.J.; Groten, J.P.; Koeman, J.H.; Bladeren, P.J. van, *Toxic. Appl. Pharmac.* **1988**, *95*, 53-60.
63 Vroomen, L.H.M.; Berghmans, M.C.J; Bladeren, P.J. van; Groten, J.P.; Wissink, H.C.G.; Kuiper, H.A. *Drug Metab. Rev.* **1990**, *22*, 79-92.
64 Rico, A.G.; Burgat-Sacaze, V. *Fd Add. Contam.* **1984**, *1*, 157-161.
65 Lu, A.Y.H.; Miwa, G.T.; Wislocki, P.G. *Rev. Biochem. Toxicol.* **1988**, *9*, 1-27.
66 Hoogenboom, L.A.P.; Oorsprong, M.B.M.; Vliet, T. van; Kuiper, H.A. *Toxicology in Vitro* **1991**, *5*, 31-38
67 Ali, B.H. *Res. in Vet. Sci.* **1984**. *37*, 290-292.
68 Pang, C.Y.; Phillips, G.D.; Campbell, L.D. *Can. J. of Comp. Med.* **1982**, *46*, 321-326.
69 Simpson, C.F.; Rollinghoff, W.; Preisig, R.; Fisher, M.J. *Can. J. Comp. Med.* **1979**, *43*, 345-351.
70 Mustafa, A.I.; Ali, B.H.; Hassan, T. *Vet. Quarterly* **1987**, *8*, 295-300.
71 Holtzman, J.L. *Life Sciences* **1981**, *30*, 1-9.

72 Samuelsen, O.B.; Svardal, A.M.; Hjeltnes, B. *Bull. Eur. Ass. Fish Pathol.* **1990**, *10*, 35-36.
73 Paul, H.E.; Paul, M.F.; Kopko, F.; Bender, R.C.; Everett, G. *Endocrinology* **1953**, *53*, 585-592.
74 Paul, M.F.; Bryson, M.J.; Harrington, C. *J. Biol. Chem.* **1956**, *219*, 463-471.
75 Palm, D.; Magnus, U.; Grobecker, H.; Jonson, J. *Naunyn-Schmiedebergs Arch. Pharmak. Exp. Path.* **1967**, *256*, 281-300.
76 Stern, I.V.; Hollifield, R.D.; Wilk, S.; Buzard J.A. *J. Pharmac. Exp. Ther.* **1967**, *156*, 492-499.
77 Pettinger, W.A.; Soyangco, F.G.; Oates, J.A. *Clin. Pharmac. Ther.* **1968**, *9*, 442-447.
78 Ali, B.H.; Bartlet, A.L. *Quarterly J. Exp. Physiol.* **1982**, *67*, 69-79.
79 Hoogenboom, L.A.P.; Tomassini, O.; Oorsprong, M.B.M.; Kuiper, H.A. *Fd Chem. Toxicology* **1991**, *29*, 185-191.
80 Vroomen, L.H.M.; Berghmans, M.C.J.; Hekman, P.; Hoogenboom, L.A.P.; Kuiper, H.A. *Xenobiotica* **1987**, *17*, 1427-1435.
81 Hoogenboom, L.A.P.; Kammen, M. van; Berghmans, M.C.J.; Koeman, J.H.; Kuiper, H.A. *Fd Chem. Toxicol.* **1991**, *29*, 321-328.
82 Food and Drug Administration *Federal Register* **1976**, *41*, 19906-19921.
83 McCalla, D.R. *Environm. Mutagen.* **1983**, *5*, 745-765.
84 Vroomen, L.H.M.; Ommen, B. van; Bladeren, P.J. van; *Toxicology In Vitro* **1987**, *1*, 97-104.

RECEIVED February 28, 1992

Chapter 7

Dermal Absorption and Metabolism of Xenobiotics in Food-Producing Animals

J. E. Riviere

Cutaneous Pharmacology and Toxicology Center, College of Veterinary Medicine, North Carolina State University, Raleigh, NC 27606

Dermal absorption of agricultural chemicals and animal drugs in food-producing animals must be considered as a potential route from which tissue residues of drugs and chemicals may occur. This has been supported in studies of topical pesticide exposure in cows and sheep. Despite the many advances made in in vitro and in vivo techniques for assessing percutaneous absorption in laboratory animals and man, very little systematic attention has been focussed on food-producing animals. The only exception is the pig since it is an accepted model for human studies. The purpose of this manuscript is to overview the literature on dermal xenobiotic absorption in food-producing animals to illustrate the risk that is present, and to outline how in vitro and in vivo methods could be applied to this problem. Factors which might alter dermal absorption and metabolism of xenobiotics will be presented. This work will hopefully provide a framework for the systematic investigation of comparative dermal absorption across the diverse species of food-producing animals.

Dermal or percutaneous absorption is the process by which xenobiotics in contact with the skin are absorbed into the body, allowing them to distribute throughout the tissues of the body and exert systemic effects. In human medicine, this process is taken advantage of in the development of transdermal drug delivery systems (1). If the exposed animal is destined to enter the food chain, then dermal absorption may result in the presence of drug residues in edible tissues after slaughter. In the case of accidental exposure to agricultural chemicals, a similar situation exists. A large amount of literature and several texts (2-4) have addressed this problem in relation to predicting absorption of drugs and chemicals in man. Using in vitro and in vivo animal and human studies, the physiochemical parameters governing chemical absorption in man have been defined. The relative merits of various animal models are well known, with the pig, monkey and hairless rodent species consistently

0097–6156/92/0503–0088$06.00/0

showing strong correlations to man. However, very little work has addressed the parameters governing percutaneous absorption in food-producing animals such as the cow, sheep or goat. In fact, characteristics which would eliminate a specific animal from predicting human absorption (e.g. high hair density) may in fact be required for predicting absorption in a species such as sheep. Recent studies have also demonstrated that the skin is capable of metabolizing topically applied chemicals prior to absorption (5). Studies have been conducted in vitro in rodent and human tissue. To this author's knowledge, studies have never been conducted in food-producing animals other than the pig (6).

Definition of Principles

Percutaneous absorption is defined as the process by which a compound placed on the surface of the skin penetrates it and is absorbed into the cutaneous circulation (7). The relevant histological structure of skin is illustrated in Figure 1 (8). The primary rate-limiting barrier for absorption of most compounds is the stratum corneum with chemical permeation occurring through the intercellular lipids. This process is governed by partitioning between the vehicle and stratum corneum and then diffusion through the lipid environment. Generally, permeability correlates to high lipid/aqueous partition coefficients. However, as lipid solubility increases, the tendency for depot formation may also increase. Vehicles are important variables, with recent data suggesting that they penetrate the lipid domain and alter the permeability characteristics.

There are a number of potential fates for a compound placed on the surface of the skin (Table I). Those fates listed in **A** would result in loss of chemical from the system. This author defines compound which is handled by mechanisms **B - E** as having "penetrated" the skin while chemical proceeding to steps **D and E** as actually having been "absorbed" through the skin into the body. From the perspective of tissue residues, the situation becomes very complicated. The "absorbed" fraction is available for potential deposition into edible tissues or milk. However, the penetrated fraction not absorbed (**B and C**) may cause residues in skin and subcutaneous tissue and also result in a prolonged release of chemical into the systemic circulation. Thus compounds which have extensive binding in skin and form so-called "reservoirs" or "depots" may prolong the sojourn of chemical in the body resulting in an enhanced risk for violative residues. Knowledge of the above concepts are also important for selecting the relevant animal model, especially when in vitro systems are involved.

Overview of the Literature

The majority of the research articles identified in a search of the literature of dermal absorption in food-producing animals were concerned with studies utilizing pour-on or dipped pesticides in small groups or even individual animals. In almost all instances, specific formulations were used with quantitative data on dosimetry variables not reported. Although as will be discussed, percutaneous absorption can unequivocally be documented, data is not available in the open literature to quantitatively assess the mechanisms of absorption in different species.

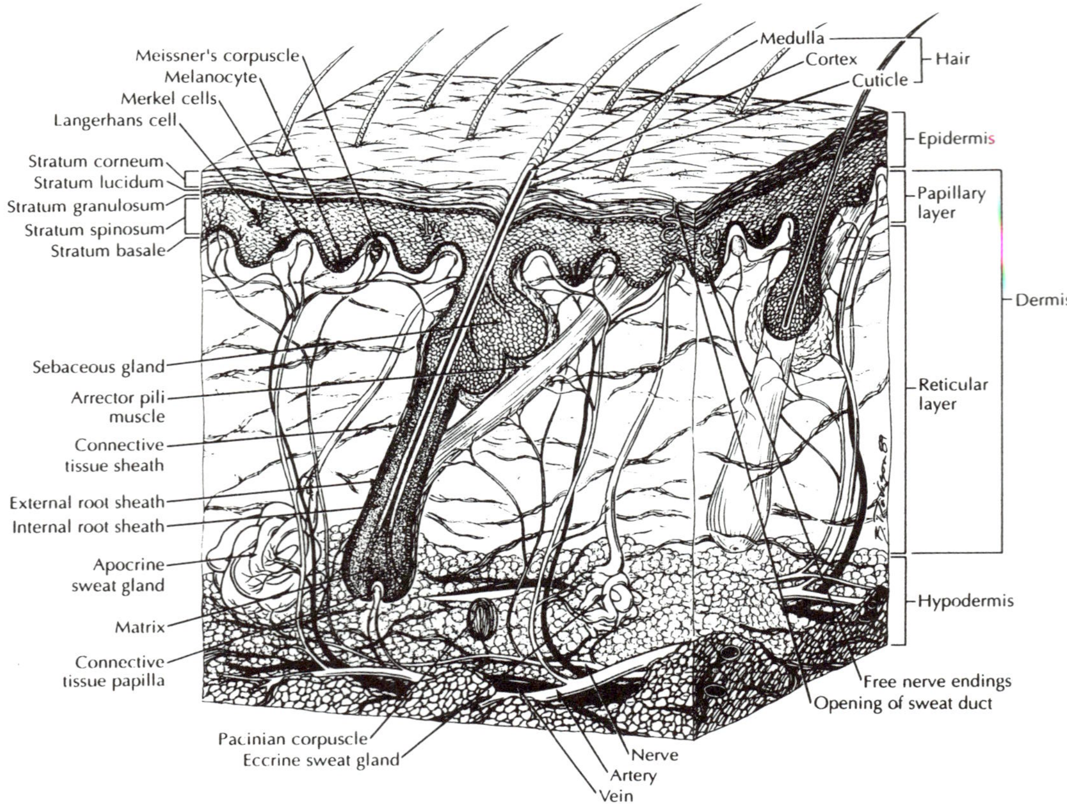

Figure 1: Schematic diagram of cutaneous microanatomy (Reproduced with permission from ref. 8. Copyright 1991 CRC Press, Inc.)

Table I. Fate of a Topically Applied Chemical

A. Dose not absorbed into stratum corneum
- Remains in the vehicle
- Binds to application device
- Evaporates if volatile
- Removed on skin by exfoliation

B. Dose penetrates the stratum corneum
- Partitions into intercellular lipids or keratinocytes forming a reservoir.

C. Dose reaches the viable epidermis
- Metabolized
- Enters basal epidermal cells

D. Dose reaches dermis
- Partitions into sub-dermal fat and forms a reservoir

E. Dose is absorbed into cutaneous capillaries and reaches systemic circulation

A number of studies have been concerned with the use of lindane dips with sheep. In one study (9), sheep plunged-dipped in a 0.0125% lindane emulsion had detectable residues in fat for up to 12 weeks after exposure. Interestingly, residues in unshorn sheep were higher (2-3 fold) and persisted longer than in similarly exposed shorn animals. In contrast, a recent study utilizing deconvolution pharmacokinetic analysis (10) demonstrated the opposite effect of fleece shearing. In this study, which is typical of the pattern seen with topical application of many lipid-soluble compounds, the whole body half-life of lindane was prolonged after spraying or dipping as compared to that seen after intravenous administration. Total absorption was 8-10% of the applied dose. This study clearly demonstrated that linear systems deconvolution analysis, previously utilized in human studies (11) was applicable to the dermal absorption of pesticides in sheep. Earlier studies with lindane (12) also demonstrated that in similarly exposed sheep and goats, ovine fat residues were greater. Studies of a similar nature conducted in lactating cows demonstrated milk residues of lindane for up to 24 days after a single administration (13).

Similar studies to the above have been reported for spraying or dipping with other pesticides, including diazinon, malathion, methoxychlor, and carbaryl (14-22). These studies generally showed detectable, albeit low, levels of residues in fat or milk. When levels were high in milk from individual animals, dilution occurring in bulk processing reduced exposure in the final product. It is difficult to draw broad conclusions from these studies because of limits on analytical sensitivity and reliability, lack of experimental controls and details. Some earlier data has also been reported on pesticide residues in the meat and eggs of exposed poultry. In an interesting study which demonstrates the toxicological hazard associated with topical exposure to pesticides (23), significant deaths (10%) occurred in a flock of 640 sheep attributed to dipping with dieldrin.

In contrast to these data, a reasonable literature base exists on the disposition of topically applied fenthion, a systemic organophosphate pesticide. Topical treatment of dairy cows with fenthion at clinically effective doses resulted in detectable but nonviolative residues in milk (24) or body fat (25). In this latter study, 90% of the residues detectable in the fat of treated cattle was parent compound. When the metabolic profile of fenthion was compared after dermal and intramuscular administration, very few differences were noted except for the time course of radiolabel clearance from the animals (26). It was not possible in these studies to study the mechanism of dermal fenthion metabolism more precisely.

There is a paucity of published data on in vitro studies of dermal penetration in food animals. In one study comparing the effects of freezing on human and cow skin (27), it is alluded to that the follicular route of penetration predominates in the cow. These authors have also written an excellent review on topical drug delivery to cattle and sheep which should be consulted for further details (28). In this work, studies on the dermal application of twenty four different substances are reviewed. Some interesting data on differential effects of dermal enhancers in different species underlines the problems inherent to interspecies extrapolations. One conclusion of this study was that "the barrier properties of sheep and cattle skin are not well understood." This situation has not improved in the intervening years.

In Vivo and In Vitro Techniques

As is evident from the above discussion, the majority of the studies reported were conducted in vivo in food animals. However, except for the deconvolution study (10), the experimental protocols used would not be acceptable to workers using animal systems to estimate human absorption. In these studies, drug is administered separately by dermal and parenteral routes, and either concentrations in the blood or excreta (urine and feces) are monitored until all of the compound is cleared from the animal. The ratio of recoveries after dermal and parenteral application is a measure of the bioavailability of the compound. A study conducted in our laboratory using six radiolabelled compounds in pigs illustrates these techniques (29). For four of the six compounds studied, the fraction of total radiolabel excreted by the fecal route was greater after topical administration than after intravenous dosing. This suggests that the systemic disposition of a topically applied compound cannot be predicted solely from a knowledge of parenteral disposition. One component of this difference may be due to "first pass" cutaneous metabolism. Elimination half-lives were generally longer after dermal dosing, an observation consistent with the hypothesis that formation of cutaneous depots results in prolonged elimination from the body. This is a potential concern for residue formation. Although in vivo studies are useful to directly assess the extent of absorption and kinetics of tissue residues, the techniques are laborious and time-consuming and the results may often be confounded by systemic metabolism of the compound. Studies addressing the mechanism of absorption or metabolism are difficult to design. Studies with very toxic pesticides often cannot be conducted.

In vitro strategies utilize full thickness or dermatomed skin slices in static or flow-through diffusion cell systems, the full details of which are presented in some excellent texts (2,3). These experiments thus assess the absorption and penetration of chemical through the epidermis and various layers of dermis. A potential pitfall is with lipid-soluble compounds that partition into the dermal tissue and do not enter the perfusate. In these situations, monitoring of absorptive flux will underestimate the true extent of percutaneous absorption. However, if the dermal tissue is assayed for compound at the termination of an experiment, a better estimate of systemic absorption may be obtained. These techniques have been widely used by many investigators. Our laboratory has studied the percutaneous absorption of parathion under different environmental conditions in an in vitro flow-through diffusion cell system using dermatomed pig skin (30). These studies clearly demonstrated that elevated temperature or relative humidity enhanced the dermal absorption of parathion. Recent advancements in in vitro flow-through cell techniques have allowed these systems to be used to study dermal metabolism (31). These systems would be very well suited to investigating the comparative absorption and dermal metabolism of agricultural chemicals in food-producing animals.

The next level of in vitro system development incorporates a functional cutaneous microcirculation into the system by utilizing a perfused vascularized skin flap model. Our laboratory developed such a system using porcine skin to study the absorption and penetration of topically applied drugs and chemicals (6,32-37). The advantage of this system, whose predictions correlate well with in vivo data, is that the mechanisms of percutaneous absorption may be modelled independent of whether the rate-limiting step in absorption is related to diffusion through the stratum corneum, metabolism, release from a cutaneous depot or vascular uptake. The venous efflux profile of topically applied carbaryl demonstrating the nature of the data produced is shown in Figure 2.

The main advantage of this approach is that pharmacokinetic models can be readily fitted to the experimental data allowing for a better understanding of the mechanisms involved. For example, the model in Figure 3 has been utilized to study the absorption of many compounds. If the compound tends to form a cutaneous depot, it will be detected by the presence of compartment 3. This could be secondary to partitioning or binding phenomenon in skin. The cutaneous efflux profile predicted by this model then serves as the input profile into a systemic pharmacokinetic model allowing for the effect of depot formation (or cutaneous metabolism) to be studied. It also allows data obtained from independent parenteral studies to be used in making risk assessments after dermal exposure. In this mode, the skin flap can best be viewed as a "living infusion pump." Since these studies have been conducted in porcine skin, direct extrapolation to swine is obvious. We have also created isolated perfused skin flap preparations in horses and there is no inherent reason preventing development of a model in cows, goats or sheep.

Integrated Approach to the Problem

As can be appreciated from the above discussion, dermal exposure to pesticides may result in the presence of tissue residues in food-producing animals. There exist powerful new in vitro and in vivo approaches to study this problem systematically and gain an understanding of the mechanisms of percutaneous absorption in a number of food animal species. The main question which must be posed is whether dermal absorption in cows, sheep or goats is controlled by mechanisms which are fundamentally different from those operative in other animal systems. As alluded to earlier, differences in hair density, glandular secretions (e.g. lanolin in sheep), epidermal structure and metabolic pathways may prevent inter-species extrapolations. This would be similar to the difficulty encountered in predicting human absorption from studies conducted in haired rodents such as mice and rats. Paradoxically, these species may be useful to predict absorption in species such as sheep. Our laboratory has recently completed an interspecies analysis of the comparative histology and blood flow between nine species at different body sites (38), demonstrating significant differences which could affect dermal absorption.

Based upon working principles learned from dermal absorption studies conducted to date, it is this author's contention that one should be concerned about tissue residues under the following conditions:

(i) very high dermal bioavailability

(ii) formation of cutaneous depots (e.g. significant penetration coupled with minimal absorption)

(iii) extensive first pass cutaneous metabolism

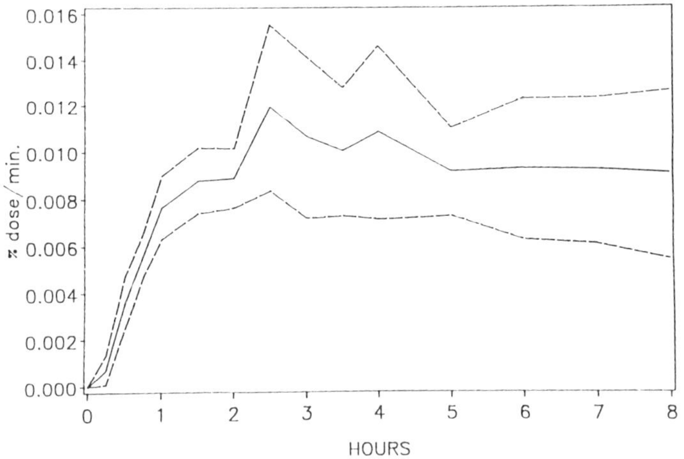

Figure 2: Cutaneous efflux profile of topically applied carbaryl (40 mcg/cm^2) in the isolated perfused porcine skin flap. Mean ± S.E., n=4

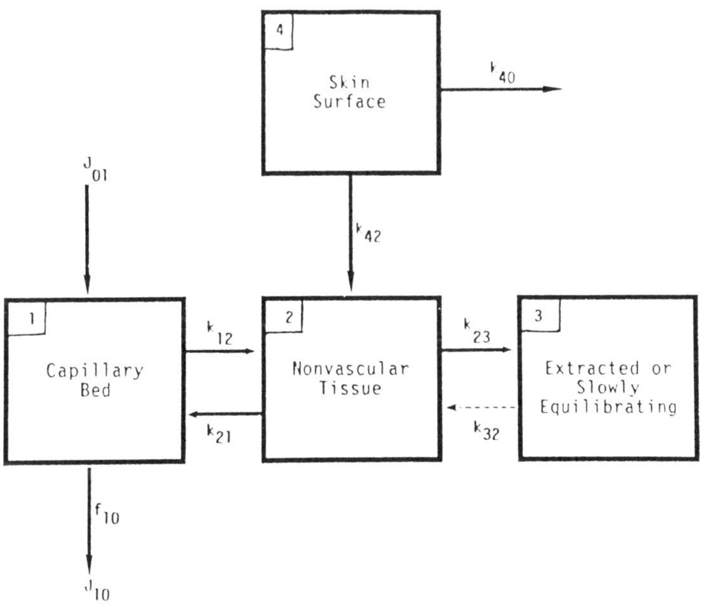

Figure 3: Pharmacokinetic model used to study absorption and disposition of a topically applied compound in the skin. Compartment 3 may have implications in predicting tissue residues.

On the other hand, if the compound only slightly penetrates the skin, the risk of tissue residues should be minimal. Under all circumstances, residues at the application site must be considered separately from tissue residues resulting from systemic distribution. If depots form, the nature of the binding (reversible or irreversible-covalent) must be assessed. Although this list seems formidable, it is amenable to straightforward experimental procedures.

A rational approach to assess these issues would be to evaluate a series of chemically distinct compounds using flow-through in vitro cell systems with pig, cow, goat and sheep skin. Ideally these studies should be conducted at more than a single dose and with three different vehicle systems. From this initial work, relative permeability constants could be calculated and the extent of cutaneous metabolism determined. The effects of vehicles and dose linearity would also be assessed. If significant species differences existed, they should become evident with these studies. The predilection to form cutaneous depots could also be identified. Similar studies could then be conducted in perfused animal skin flap models to determine quantitative cutaneous efflux profiles useful for integrating into systemic pharmacokinetic models. Such a database would go a long way in assessing the risk of tissue residues after dermal exposure.

In conclusion, dermal exposure of chemicals in food-producing animals may result in the presence of tissue residues. For topical products, formulations which minimize dermal penetration should also minimize residues. However, for topically applied, systemically active drugs, tissue residues may occur. Our knowledge of the penetration characteristics of chemical in food-producing animals is lacking. In vitro studies designed to probe the mechanism of dermal absorption in these species must be conducted for rational decisions to be made.

Literature Cited

1. Transdermal Drug Delivery; Hadgraft, J.; Guy, R.H., Eds.; Marcel Dekker: New York City, New York, **1989**.
2. In Vitro Percutaneous Absorption: Principles, Fundamentals, and Applications; Bronaugh, R.L.; Maibach, H.I., Eds.; CRC Press: Boca Raton, Florida, **1991**.
3. Methods for Skin Absorption; Kemppainen, B.W.; Reifenrath, W.G., Eds.; CRC Press: Boca Raton, Florida, **1990**.
4. Dermal Exposure Related to Pesticide Use; Honeycutt, R.C.; Zweig, G.; Ragsdale, N.N., Eds.; American Chemical Society: Washington, D.C., **1985**.
5. Kao, J.; Carver, M.P. Drug Metab. Rev. **1990**, 22, 363-410.
6. Carver, M.P.; Levi, P.E.; Riviere, J.E. Pest. Biochem. Physiol. **1990**, 38, 245-254.
7. Riviere, J.E. Cosmet. Toilet. **1990**, 105, 85-93.
8. Monteiro-Riviere, N.A. In Fundamentals and Methods of Dermal and Ocular Toxicology; Hobson, D.W., Ed.; CRC Press: Boca Raton, Florida, **1991**, Chapter 1.
9. Collett, J.N.; Harrison, D.L. N.Z.J. Agric. Res. **1963**, 6, 39-42.

10. Dagorn, M.; Guillot, P.; Sanders, P.; Laurentie, M.; Toritoin, P.L. Am. J. Vet. Res. **1990**, 51, 1993-2000.
11. Fisher, H.L.; Most, B.; Hall, L.L. J. Appl. Toxicol. **1985**, 5, 163-177.
12. Jackson, J.B.; Ivey, M.C.; Roberts, R.H.; Radeleff, R.D. J. Econ. Entomol. **1959**, 52, 1031-1032.
13. Oehler, D.D.; Hogan, B.F.; Graham, O.H. J. Econ. Entomol. **1970**, 63, 1467-1468.
14. Bourne, J.R.; Arthur, B.W. J. Econ. Entomol. **1967**, 60, 402-405.
15. Camp, H.B.; Buttram, J.R.; Hays, K.L.; Arthur, B.W. J. Econ. Entomol. **1963**, 56, 402-404.
16. Cheng, T.H.; Frear, D.E.H.; Enos, H.F. J. Econ. Entomol. **1958**, 51, 618-623.
17. DeFoliart, G.R.; Willett, D.N. J. Econ. Entomol. **1961**, 54, 871-873.
18. Eheart, J.F.; Turner, E.C.; Dickinson, J. J. Econ. Entomol. **1962**, 55, 504-505.
19. Eschle, J.L.; Oehler, D.D.; Hogan, B.F.; Ivey, M.C. J. Econ. Entomol. **1973**, 66, 290-291.
20. Ivey, M.C.; Eschle, J.L.; Hogan, B.F. J. Econ. Entomol. **1971**, 64, 320-321.
21. March, R.B.; Metcalf, R.L.; Fukoto, T.R.; Gunther, F.A. J. Econ. Entomol. **1956**, 49, 679-682.
22. Oehler, D.D.; Eschle, J.L.; Miller, J.A.; Claborn, H.V.; Ivey, M.C. J. Econ. Entomol. **1969**, 62, 1481-1483.
23. Glastonbury, J.R.W.; Walker, R.I.; Kennedy, D.J.; Gill, P.A.; McDougall, K.W.; Sharrock, A.G. Aust. Vet. J. **1987**, 64, 145-148.
24. O'Keeffe, M.; Eades, J.F.; Strickland, K.L. J. Sci. Food Agric. **1983**, 34, 192-197.
25. Wright, F.C.; Riner, J.C. J. Agric. Food Chem. **1979**, 27, 576-577.
26. Knowles, C.O.; Arthur, B.W. J. Econ. Entomol. **1966**, 59, 1346-1352.
27. Pitman, I.H.; Rostas, S.J. J. Pharm. Sci. **1982**, 71, 427-430.
28. Pitman, I.H.; Rostas, S.J. J. Pharm. Sci. **1981**, 70, 1181-1194.
29. Carver, M.P.; Riviere, J.E. Fundam. Appl. Toxicol. **1989**, 13, 714-722.
30. Chang, S.K.; Riviere, J.E. Fundam. Appl. Toxicol. **1991**, 17, 494-504.
31. Bronaugh, R.L; Stewart, R.F; Storm, J.E. Toxicol. Appl. Pharmacol. **1989**, 99, 534-543.
32. Riviere, J.E; Bowman, K.F; Monteiro-Riviere, N.A; Carver, M.P; Dix, L.P. Fundam. Appl. Toxicol. **1986**, 7, 444-453.
33. Riviere, J.E; Bowman, K.F; Monteiro-Riviere, N.A. Brit. J. Dermatol. **1987**, 116, 739-741.
34. Carver, M.P; Williams, P.L; Riviere, J.E. Toxicol. Appl. Pharmacol. **1989**, 97, 324-337.
35. Williams, P.L.; Riviere, J.E. J. Pharm. Sci. **1990**, 79, 305-311.
36. Riviere, J.E; Sage, B.S; Williams, P.L. J. Pharm. Sci. **1991**, 80, 615-620.
37. Riviere, J.E.; Monteiro-Riviere, N.A. Crit. Rev. Toxicol. **1991**, 21, 329-344.
38. Monteiro-Riviere, N.A; Bristol, D.G; Manning, T.O; Rogers, R.A.; Riviere, J.E. J. Invest. Dermatol. **1990**, 95, 582-586.

RECEIVED March 13, 1992

Chapter 8

Drug Pharmacokinetics and Metabolism in Food-Producing Fish and Crustaceans

Methods and Examples

Kevin M. Kleinow[1], Margaret O. James[2], and John J. Lech[3]

[1]Department of Pharmacology, School of Veterinary Medicine, Louisiana State University, Baton Rouge, LA 70803

[2]Department of Medicinal Chemistry, College of Pharmacy, University of Florida, Gainesville, FL 32610

[3]Department of Pharmacology, Medical College of Wisconsin, Milwaukee, WI 53226

The worldwide demand for fish and shellfish as a major item of the human diet has led to an increase in the use of aquaculture. Disease is often a problem in culturing aquatic species, and has led to the perceived need for veterinary drugs for disease control. In order to use these drugs efficiently and safely, it is necessary to understand how they are handled by the fish and shellfish in which they will be used. The design of pharmacokinetic studies in fish and shellfish must take into account the methods for obtaining samples of blood and excreta in the aquatic environment without unduly stressing the animal. Many quantitative and some qualitative differences in drug biotransformation exist between fish, shellfish, and more commonly encountered veterinary and laboratory animals. These differences must be considered in appropriately designing *in vivo* and *in vitro* studies. Studies of the fate and residue dynamics of several drugs and model xenobiotics, including oxolinic acid, sulfadimethoxine, ormetoprim, erythromycin, phenol and benzoic acid have shown that of the many factors that influence the metabolism and elimination of drugs, species is perhaps the most important. These findings underline the need for conducting studies in the target species.

While the biotransformation of xenobiotic chemicals and drugs has been systematically studied in mammalian species for many years, similar investigations in fish species lagged behind mammalian research, and it is only in the last 5 years or so that the importance of biotransformation in fish has been recognized. *In vitro* studies utilizing fish tissue fractions and model substrates have demonstrated phase I and phase II biotransformation reactions, although the concentration of specific enzymes varied among species (*1-4*). These studies indicated that fish have the capacity to biotransform a variety of substrates, but, generally, the rates observed were lower than those in many mammalian species.

0097–6156/92/0503–0098$09.25/0

Monooxygenase activity associated with isolated hepatic microsomes of several species of fish has been demonstrated to be sensitive to inhibition by carbon monoxide and responsive to mammalian monooxygenase modulators such as α-naphthoflavone and metyrapone (*5-7*). Sufficient evidence exists to indicate that phase I reactions demonstrated in fish *in vitro* are catalyzed by a number of forms of microsomal cytochrome P450, which are similar to those found in mammals (*5,8,9*). Although the purification and characterization of cytochrome P450 forms from fish tissues have not approached the depth observed with mammalian systems, recent studies indicate that the substrate specificities are similar to those seen in mammalian systems (*10*). That both benzphetamine-N-demethylase and ethylmorphine-N-demethylase activities have been demonstrated in hepatic microsomes of trout suggests that a system comparable to the cytochrome P450 II family is present, but evidence indicates that this cytochrome family in fish appears to be refractory to induction by drugs and chemicals which produce phenobarbital-type induction in mammals (*11,12*). The presence of constitutive, noninducible form(s) of P450 in rainbow trout liver has recently been confirmed by chromatographic, immunological and molecular probe techniques (*10,13*). On the other hand, monooxygenase activity toward ethoxyresorufin (de-ethylation), a highly specific substrate for the polycyclic aromatic hydrocarbon-inducible cytochrome P4501A1 has been shown to be present in hepatic microsomes from several species of fish (*11,13*). This latter form of P450 is inducible by 3-methylcholanthrene and other inducers of this class, including polycyclic aromatic hydrocarbons and several co-planar polychlorinated biphenyl congeners (*11,13,14*). Other phase I and phase II reactions have been demonstrated *in vitro* with tissue preparations from various fish species. It is clear that glucuronyl transferase, sulfotransferase, glutathione-S-transferase, and epoxide hydrolase activities are, at least qualitatively, similar to those found in mammals (*6*). Although investigations *in vitro* have been useful in characterizing the enzymes involved in the biotransformation of chemicals in fish, one can realize that an area of equal or greater importance is the functional significance of these biotransformation reactions *in vivo*. It is noteworthy that almost every biotransformation reaction (phase I and phase II) that has been described *in vitro* in fish has occurred *in vivo* during exposure of fish to various chemicals and drugs, and metabolites arising from biotransformation reactions have been identified in fish tissues including liver, bile, and urine (*15,16*).

Early *in vitro* studies of cytochrome P450-dependent monooxygenase activities in decapod crustacea showed that although cytochrome P450 was present, little or no monooxygenase activity could be measured in hepatopancreas microsomes. The low activity was due to the presence of inhibitors of monooxygenase activity that were released during the preparation of microsomes from this source (*17*). It is now clear that the crustacean hepatopancreas is a major site of xenobiotic oxidation *in vivo*, although crustacea vary considerably in their capacity for drug monooxygenation, with the Maine lobster *Homarus americanus* having a very low monooxygenase ability (James, M.O. *Marine*

Environ. Res., submitted). Some monooxygenase activity has been found in microsomes prepared from the stomach and antennal gland of crab and crayfish species (*18,19,20*). Of the other phase I enzymes, high epoxide hydrolase has been found in hepatopancreas (*21*), but little information is available on ester hydrolase activity in crustacea.

Glucuronidation, a major pathway of drug conjugation in vertebrate (terrestrial and aquatic) species, has not yet been unequivocally found in crustaceans. Glucuronide conjugates have not been positively identified as metabolites in crustacean tissues or excreta, and no UDP-glucuronosyl transferase activity has been found in microsomes from hepatopancreas or antennal gland of the crayfish, the lobster or the spiny lobster, *Panulirus argus* (*22*). Instead, crustacea conjugate hydroxy groups with glucose, and the conjugation is catalyzed by UDP-glucosyl transferase (*23*). Like fish, crustaceans conjugate carboxylic acids with taurine, not glycine (*24,25*). There have been few studies of drug acetylation in crustacea. Studies of sulfadimethoxine disposition in the lobster have shown that N-acetylsulfadimethoxine, a major vertebrate metabolite, is formed in very small amounts, if at all, by the lobster (*26*). The other major pathways of drug conjugation, namely sulfation and glutathione conjugation, are found in crustaceans (*22*). The information presented in Table I illustrates a partial listing of biotransformation reactions which have been demonstrated in various aquatic species *in vivo* with specific drugs and chemicals.

From a functional point of view, biotransformation reactions can significantly influence the biological properties of drugs and chemicals, depending on the qualitative nature of the reaction and the rate at which it occurs *in vivo* (*15,16*). Table II illustrates that several important pharmacokinetic parameters may be influenced by biotransformation reactions. Whereas these important areas have not been extensively studied in aquatic species, there exists ample evidence that the biotransformation of certain chemicals and drugs including antibiotics in fish may be critical to several and possibly all, of these parameters. While the intimate details of the biotransformation of many chemicals and drugs in these species remain to be elucidated, apparently the reactions leading to metabolites of commonly used antibiotics occur in these species *in vivo*, based on metabolism studies and pharmacokinetic studies (*26-27*) (*Kleinow, K.M. Can. J. Fish Aq. Sci.*, in press).

An understanding of metabolism and kinetics is pertinent to the overall goal of this Symposium, inasmuch as the rate and pathway of the biotransformations in a given species are important determinants of the dose-response relationship of the therapeutic agent and its elimination kinetics and terminal residues. Clearly, the current literature reveals that biotransformation reactions play a role in the overall behavior of chemicals and drugs in aquatic species and that the manipulation of the rates of these reactions by inducers or inhibitors of biotransformation can significantly affect important kinetic and therapeutic parameters, such as steady-state tissue levels, persistence, and metabolite profiles.

Table I. **Biotransformation Reactions Demonstrated in Fish and Crustacea**

Phase I	*Phase II*	*Species*	*Chemical*
O-Dealkylation		Fathead minnow	*p*-Nitrophenylethers
		Rainbow trout	Pentachloroanisole
			Fenitrothion
N-Dealkylation		Carp	Diethylnitrosamine
		Spiny lobster, Lobster	Benzphetamine
			Erythromycin
Oxidation		Mudsucker, Sculpin	Naphthalene
			Benzo[a]pyrene
		Coho salmon	Naphthalene
		Rainbow trout	Methylnaphthalene
		Carp	Rotenone
		Mosquito fish	Aldrin, Dieldrin,
		Spiny lobster, Blue crab	Benzo[a]pyrene,
Hydrolysis		Catfish, Bluegill	2,4-Dichloro- phenoxyacetic acid esters
		Rainbow trout	Diethylhexylphthalate
		Pinfish	Malathion
		Blue crab	Fenitrothion
		Mosquito fish	Parathion
Hydration		Flounder, Sheepshead	Styrene oxide, Benzo[a]pyrene 4,5-oxide,
		Spiny lobster, Blue crab	Octene oxide
	Acetylation	Dogfish shark,	Ethyl-*m*-aminobenzoate
		Rainbow trout	Ethyl-*m*-aminobenzoate
			Sulfadimethoxine
		Catfish	Sulfadimethoxine
	Glutathione conjugation	Carp	Molinate
	Taurine conjugation	Flounder	2,4-Dichloro- phenoxyacetic acid
		Catfish, Trout	Benzoic acid
	Sulfate conjugation	Goldfish	Pentachlorophenol
		Lobster	Phenol
	Glucuronide conjugation	Rainbow trout	Pentachlorophenol
		Rainbow trout	3-Trifluoromethyl- 4-nitrophenol
		Goldfish	Pentachlorophenol
	Glucoside conjugation	Spiny lobster	4-Methylumbelliferone
			3-Hydroxy- benzo[a]pyrene
		Lobster	4-Nitrophenol, Phenol Naphthol

Table II. Determinants of the Concentration of Active Drug in Target Tissues and the Elimination of Drug Residues

1. Uptake
   - Route of administration
   - Blood flow
   - Lipid solubility and charge
2. Rate of metabolism to active drug
3. Rate of inactivation of active drug
4. Rate of renal, branchial and gut excretion of parent compound and metabolites

Methods for the Study of Drug Disposition in Fishes

Specific methods for the study of biotransformation and pharmacokinetics of drugs in fish have been developed to address the unique characteristics of this group of vertebrates. While similar types of studies can be performed in fish as in their mammalian counterparts, their success require approaches which deal with an animal's existence in water, provide controlled water quality, circumvent stress related problems and recognize the inherent differences in anatomy and physiology. Especially problematic is the diversity encountered among the 20,000 members of this group of animals. One species of fish can differ markedly from another in terms of life history, water quality and temperature requirements, internal anatomy, physiology and behavioral considerations. These differences become paramount when addressing these animals experimentally. The proper husbandry and handling of the fish is a major determinant of success even before a study begins. Failure to address such issues may lead to unsuccessful experiments or data of questionable value.

Anatomical and Physiological Considerations. Fish possess a large number of unique features which distinguish them both structurally and functionally from other vertebrates. Many of these biological characteristics are germane to xenobiotic studies.

Gills. The gill is one of the most versatile organs in fish. Critical functions such as acid- base balance, electrolyte balance, and nitrogenous waste product elimination are often overshadowed by the role of the gills in gas exchange. Gills are also important avenues for xenobiotic absorption and elimination. This is especially true for highly polar compounds which are non-ionized and moderately lipid soluble. The former is dependent on the pH of the transport media (blood, water) and the pKa of the drug. Mechanisms which serve to maximize extraction of oxygen from a relatively oxygen deficient water environment (20% of that in air) such as counter current flow of blood and water, thin diffusion barriers (2 cells thick) and, lamellar recruitment (surface area) also serve to promote xenobiotic transfer. Water quality alterations in pH (*28*) as well as hypoxia induced respiratory adjustments (*29*) have been shown to alter the absorption and disposition of waterborne xenobiotics.

Circulation and Blood. Blood volume of many fish species has been reported to lie within the range of 3-7% of total body weight. Therefore, fish size is a major factor in protocol design. This is especially true for pharmacokinetic studies where no more than 15% of the total blood volume should be taken by sampling. Blood pH in a variety of fish species has been shown to be higher than typically reported in mammals. In the trout for example, pH values average about 8.0, however, if stressed by handling for a duration as short as 30 seconds values average around 7.5 (30). Such factors may influence drug ionization.

Total plasma protein content differs in fish as compared to mammals. Total plasma protein in the trout and flounder is approximately one half that of mammals such as dogs and cats. (Kleinow, K.M., Louisiana State University, unpublished data). For many compounds protein binding is considerably lower in fish than their mammalian counterparts (*27,31,32*).

Importantly from an experimental standpoint, the circulatory system of fish is unique structurally and functionally. Structurally, the membranous nature of the vasculature make for a friable high capacitance system under low pressure. Low blood flows result in somewhat longer distributional phases for many drugs. Processes such as heart rate and stroke volume which influence drug distribution are themselves influenced by external factors such as temperature and stress.

Hepatic Function. The liver of fish can change dramatically in regards to weight and chemical composition. These changes are often based upon alterations in glycogen and lipid content. Seasonal gonadal maturation (*33*), nutritional considerations such as time from last feeding (*34*), and response to stress may influence these parameters. Experimentally, the resulting differences can alter the functional mass used for determination of hepatic uptake, distribution and clearance of drugs.

Unlike mammals, fish do not appear to exhibit a defined relationship between portal spaces and terminal hepatic venules. This apparent lack of organization also has been reported on the biochemical level. Enzymatically, heterogenous zones of hepatocytes as found in mammals are not evident in fish (*35,36*). In addition hepatic perfusion is poor in fish with rates 1/2 to 1/4 of those found in mammals. These features may be important in regards to metabolism and toxicity of xenobiotics.

Fish are extremely good biliary concentrators of xenobiotics. This ability has prompted the suggestion that bile can be used as a biomonitor for environmental exposure of xenobiotics (*37*). Molecular weight and polarity concerns for biliary elimination are basically similar to mammals. Bile formation in fish, however, is nearly 50 times slower than mammals (*38,39*).

Renal Function. The kidney of most fish species is primarily involved in hematopoiesis (head kidney) and, osmoregulation (trunk kidney). Unlike mammals the kidney is not active to any appreciable degree in the elimination of nitrogenous waste products. Freshwater fish are hypertonic relative to the water. The continual osmotic uptake of water is balanced by production of large amounts (2-4 ml/kg/hr) of dilute (Urine Specific Gravity 1.001) urine. Saltwater fish on the other hand are hypotonic relative to their environment resulting in body water loss. In response, marine species produce much smaller volumes of urine. Large differences exist between divergent fish species in regards to nephron structure to facilitate these functional responses. These features may influence the renal contribution to xenobiotic disposition.

Digestive Function. Unlike freshwater species, marine species drink appreciable quantities of water to maintain their hydration status potentially providing waterborne xenobiotics access to uptake by the gastrointestinal tract. Regional pH in of the gastrointestinal tract may vary considerably between fish species. For example, the pH of the stomach may be much more alkaline in herbaceous fish than carnivores.

Design of Pharmacokinetic and Disposition Studies

It is important that the experimental design takes into account the close relationship of fish and their water environment. The data collected is only as good as the water quality, control of water quality throughout the experiment, the match of the water quality with the species under investigation and the acclimation procedure to the desired experimental conditions. Much like the description of analytical methods, water quality conditions should be reported to ensure reproducibility. Of methods employed for studies in fish, these are the most often overlooked.

Temperature, pH, hardness, dissolved oxygen, ammonia, total organic carbon and alkalinity are the water quality parameters which should be regulated and monitored. In addition, exposure water should be filtered and polished over activated carbon to remove potential contaminating organic compounds. Aside from preferred water quality conditions which differ between species, fish can adapt to more divergent less desirable conditions provided the rate of change is minimized. Fish under such conditions may be physiologically different. For example catfish at 11°C have, different fatty acid profiles, isozymes for muscle activity, heart rates, bile formation rates, feed requirements and growth potential than animals at 24°C (preferred). As described later in this chapter, temperature has a profound effect on drug disposition. While all the water quality parameters are contributing factors for the health of the animals a number of these items also determine the availability, chemical form and toxicity of xenobiotics. This is particularly true for waterborne exposures where the water quality characteristics modify drug disposition much like internal modifiers (protein binding, biotransformation, compartmental pH) intervene in a similar fashion.

Water in a true sense is an outward extension of the fluid space of the aquatic animal.

While protocol design in terms of collection of a quality data set (numbers of animals, use of radiolabel, analytical concerns) does not inherently differ from other species, the way the fish are acquired, acclimated, housed, exposed and sampled can influence the variability and data quality. In general, more experimental control is required for studies with fish. This may be a confounding concern as inherent difficulties may exist in obtaining healthy, genetically pure, easily sexed and experimentally acceptable animals.

Sampling and Administration Procedures. Fishes existence in water presents methodological necessities in terms of drug administration and sample collection. In practical terms dosing and taking samples from fish out of water is much like taking the same from rats underwater. This necessitates at the very least abbreviated procedural steps and preferably techniques which circumvent the problems generated by land-based sampling. These problems are most acute in pharmacokinetic studies where multiple samples may be necessary. Removing animals from the water and in fact just the netting process dramatically alters the physiology of these animals. The effect of such stress is often overt morbidity and mortality especially on extended time courses and perhaps more disconcertingly can produce subtle effects that influence the data collected.

Intravascular Administration and Blood Collection. Drug administration and blood collection can be accomplished by syringe, vac-u-tainer and cannula in most fish species. The major determinants for method selection are anatomical, size, and experimental design considerations.

The most universally used site for venipuncture in fish is the caudal vein accessible behind the anal fin in the caudal peduncle. This vessel located along the hemal arch in the ventral portion of the vertebral column can be approached laterally at a level slightly below the lateral line or by a dorsally oriented venipuncture originating on the mid ventral aspect of the caudal peduncle (Figure 1). In both approaches the ventral vertebral column should be used as a point of reference.

An alternative site for venipuncture is the cardinal sinus which in most species lies deep to the dermal sheet comprising the caudal extent of the branchial chamber. This route is usually more difficult to hit as the landmarks are not defined by boney structures. In certain fish species with a dorsal ventral flattening like skates, this is a preferred site for venipuncture.

Cardiac puncture can be used in larger fish. Trauma in conjunction with the small size of the heart make this a less desirable alternative. The heart in most species can be approached from the caudal border of the pectoral girdle.

Other blood collection strategies have been employed including collection from the orbital sinus, and the severed caudal peduncle by capillary tube. These procedures are usually used with small animals (< 5g) and the latter by its nature is terminal.

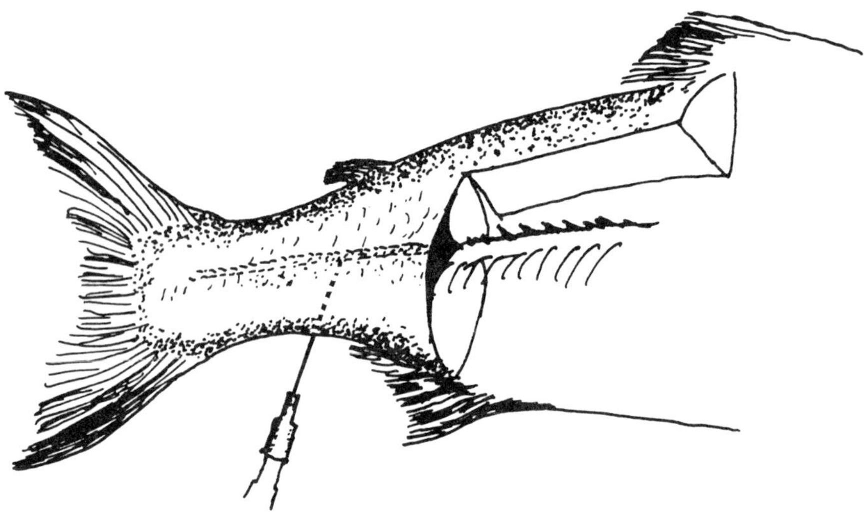

Figure 1. Schematic diagram of caudal vein venipuncture.

Cannulation procedures provide a means to administer drugs and collect samples without removing the animals from the water. Recent studies with the antibiotic oxolinic acid in catfish have compared the pharmacokinetics of the drug when administered and samples collected by netted land based caudal peduncular venipuncture with those following dorsal aorta cannulation of free swimming fish (*40*). Venipuncture resulted in much greater whole body clearances and shorter elimination half lives ($t_{½β}$) than those obtained by aortic cannulation. In addition, the redistribution phase ($t_{½α}$) was longer and more variable for the non-cannulated animals. These findings have been attributed to the stress response of handling and backflow along the venipuncture sites respectively.

Cannulation techniques employ a variety of cannulation sites including the dorsal aorta, ventral aorta and caudal peduncular vein. The ventral aorta cannulation is usually reserved for pre-branchial blood sample collection. It is not commonly used in drug studies as it is more difficult to install and use. Caudal peduncular cannulation is useful for animals such as flounder which have a ramifying aortic vasculature which would otherwise preclude cannulation. Dorsal aorta cannulations, the most common technique used, can be applied to submerged restrained, restrained submerged spinally transected (*41*) and free swimming fish (*27,31,42*). The dorsal aorta cannulation procedures are described in detail elsewhere (*43*). While variations on the theme exist, in general, the aortic cannula (PE 20-100) is inserted into the aorta along the dorsal midline of the mouth between the first to third branchial arch using an appropriately sized stylet. The cannula placement is dependent on species. After positioning the cannula may be sutured to the palatine mucosa or held in place by connection to a fitting exteriorizing the cannula dorsally through the cartilage rostral to the nares. Once available on the dorsal surface the free end of the cannula is accessible for sampling outside of the restraining chamber. For animals intended for free swimming preparations the cannula can be secured to the dorsal fin and the free end threaded through a central standpipe of an elevated circular chamber. Such an arrangement allows for a gravity dependent cannula tending system for free swimming fish.

A variety of anticoagulants can be successfully utilized for blood collection in fish. Care must be exercised in preparations in which anticoagulants are utilized to keep cannulas patent. Excess anticoagulant in conjunction with immersion of the fish in essentially a solvent environment may result in uncontrollable bleeding.

Oral Administration. Drugs may be orally administered to fish in the diet or by gavage. The ad lib feeding of medicated diets is generally undesirable for experimental purposes. Social requirements for feeding in some species, variation of feeding with external stimuli, and the large variation in amounts consumed between animals make drug studies based on ad lib feeding difficult to interpret at best. Most metabolic and pharmacokinetic studies require dosing by gavage. This technique can reliably deliver set amounts of drug in a choice of carrier vehicles (saline to food). Anatomically, intubation in fish is easier than in mammals as the esophagus is large and distensible. There is no trachea to contend with, however, aspiration onto the gills is possible upon emesis. It is important to gauge the volume of the carrier vehicle to the size of the fish and to limit the air placed in the gastrointestinal tract. Excesses in either area can evoke a well developed emetic response. Animals need to be observed individually for emesis at least five minutes before inclusion into a drug study.

Delivery techniques useful in our hands include gavage by syringe with a PE 260 extension or by an indwelling stomach tube (*40*). The latter is particularly useful for multiple dosing or where handling is undesirable.

Urinary Catheter. An appreciable amount of drug as parent or metabolites may be excreted in the urine of fish. A number of studies have demonstrated that glucuronide, sulfate and taurine conjugates are excreted by the fish kidney as a result of anion/cation carrier-mediated mechanisms (*44,45*). Urine has been collected for xenobiotic studies in a variety of fish species including flounder (*46*), dogfish shark (*47*), rainbow trout (Kleinow, K.M. *Can. J. Fish Aq. Sci.*, in press) and catfish (*48*). In large measure urine production is greater in freshwater fish as compared to marine species.

Urinary catheterization for all species generally employs a PE tubing inserted up the caudal aspect of the urogenital sinus into the urinary bladder. Proximally, the tubing is secured to the urogenital papilla and to the anal fin. Distally the tubing may connect to a collection bladder attached to the fish or continue out of the holding chamber for collection in a land based vessel. The method selected is based on convenience, the duration of collection (urine volume), the interval or aliquot size required (rate considerations) and if handling would influence the desired data set. Both experimental protocols may be used in restrained or free swimming fish.

Branchial Elimination. Elimination of xenobiotics across the gill is highly dependent upon the lipophilicity and charge of the compound at physiological pH. Compounds, like the fish anesthetic MS-222 which are moderately lipophilic and neutrally charged, are readily eliminated across the gill. Elimination and biotransformation products originating from the gill can be collected by two methods. The barrier technique separates the head and gills from the remainder of the fish by use of a divided chamber and a latex collar. This technique, which may be run static or flow-through, requires rigid restraint to physically separate the water which contains branchial elimination products from that of fecal and urinary origin. Other methods depend on use of urinary and fecal catheters in combination with flowing water to ensure elimination products to be of branchial origin. These animals may be loosely restrained in enclosures or can be free swimming (*40*). In all cases the holding water is run over an appropriate matrix, resin or ion exchange bed to remove compounds of interest. The major considerations for method selection is the reuptake of drug excreted, volume of water to the extracted for branchial elimination products, degree of restraint and duration of the study.

Mass Balance Studies. Mass balance studies may be performed using a combination of an aortic cannula, indwelling stomach tube (oral administration studies), urinary catheter, fecal catheter, and collection/extraction of flowing immersion water (branchial products). Unlike mass balance studies in mammals, such studies require this experimental format as the water serves as a common depot for all wastes and importantly serves as a confounding source for reuptake. These studies can be performed in restrained and free swimming animals, however, in either case require intensive effort and experimental diligence.

Intraperitoneal Injections. Drugs with a near neutral pH and appreciable solubility may be administered to fish intraperitoneally. Several anatomical considerations are germane to intraperitoneal drug administration. These relate to the presence of scales in fish and the fact that fish in general have little if any redundant tissue in the region of the epidermis, dermis and musculature. Injections should be performed with the smallest needle feasible. The injection directed under the scale in a cranial orientation along the mid ventral axis should be angled to the body wall with the fish's head in a dependent position. These steps will circumvent the scales entirely and minimize diffusional back flow of drug along the injection tract.

Intramuscular Injection. Injections can be administered to fish with similar potential for back diffusion. In addition distribution from these sites can be very slow especially at lower temperatures.

Water-borne Administration. For the waterborne route of exposure, drug solubility in water is a major concern. Use of sonication, warming, salt formation, dilution and carrier vehicles may facilitate dispersal of sparingly-soluble drugs into water-based stock solutions. Whenever possible, the drug should be added directly to dilution water without carrier solvents. Dimethylformamide, ethanol and acetone, can be used to facilitate solubilization if concentrations of the carrier do not exceed 0.01% of the final dilution water.

Drug purity and exposure concentrations should always be measured under exposure conditions. Nominal concentrations may be influenced by water quality parameters, partitioning on aquaria walls and fish biomass. Selection of exposure method must take in consideration maintenance of steady state drug delivery, environmental degradation of exposure drug, maintenance of water quality, the reuptake of excreted drug, cost and waste generation. In contrast to static systems, metered drug delivery in flow-through systems provide better control and maintenance of experimental parameters. By their nature, however, flow-through designs require more radiolabeled drug and extensive efforts in effluent drug recovery.

Anesthetics. The use of anesthetics or sedatives for restraint in sampling or cannulation procedures can, as in mammals, cause difficulties when used in conjunction with pharmacokinetic and metabolic studies. Anesthesia with MS-222, one of the most common fish anesthetics, has been shown under certain conditions to transiently inhibit hepatic microsomal monooxygenase activities (*49,50*) and alter urine flow rates (*51*). Such effects are negligible 18-24 hours post administration. These findings suggest that MS-222 anesthesia should not be used for those studies influenced by these parameters within 18 hours before or during experimentation. Alternatively, dependent upon species and previous thermal acclimation history ice water chilling can provide a safe transient, non-chemical restraint. This technique, however, is not recommended for multiple sampling or dosing procedures due to the cumulative temperature effects on factors such as residue retention.

Biotransformation and Identification of Metabolites. The study of fish biotransformation reactions *in vitro* require a number of modifications from those commonly used with mammalian tissues. In conducting studies with cellular components from fish optimization of enzymatic conditions will often require lower incubation temperatures and because of lower specific activities higher enzymatic fortification. Assay procedures such as those using reduced CO versus reduced difference spectrum for cytochrome P450 also require modification. In this case conventional P-450 assays with fish microsomes may experience detection difficulties due to hemoglobin interferences. Modifications in which both reference and sample cuvettes are equilibrated with CO followed by dithionite additions in the sample cuvette can address this issue (*52*).

Fish present few special considerations in regards to isolation and identification of metabolites. It should be noted the P-450 isozyme involvement and the resulting metabolite profiles may differ significantly from other animal species.

Examples of the Disposition of Xenobiotics in Fish

Sulfadimethoxine. Sulfadimethoxine has been shown to be efficacious in the treatment of bacterial diseases in catfish (*Ictalurus punctatus*) and rainbow trout (*Oncorhynchus mykiss*). Studies in rainbow trout have demonstrated a bi-exponential elimination of sulfadimethoxine from plasma when administered intravascularly at 42 mg/kg. The pharmacokinetic analysis of parent sulfadimethoxine demonstrated $t_{½\alpha}$, $t_{½\beta}$, V_{ss} and Cl_b values of 0.38 h, 15.9 h, 421.6 ml/kg and 21.8 ml/kg/h respectively (Kleinow, K.M. *Can. J. Fish Aq. Sci.* in press). The $t_{½\beta}$ of the terminal elimination phase was lengthened to 35.2 h following 5 daily intravenous doses of sulfadimethoxine (42mg/kg), to trout. There does not appear to be a biotransformational basis to these findings. Oral bioavailability of sodium sulfadimethoxine was shown to be approximately 63% when administered by gavage at 42 mg/kg and 50% when administered at 126 mg/kg. A significant difference in bioavailability from the sodium salt was seen when sulfadimethoxine was administered at 42 mg/kg as the non-salt-form (34%). This difference appears to be related to solubility considerations. The *in vivo* plasma protein binding was 15.8% throughout the 72 h time course studied. The protein binding appeared non-specific and non-saturable within a concentration range of 0.2 to 10 mM. The tissue distribution of radiolabel for selected tissues, and tissue half lives is shown in Figure 2. The highest concentration of sulfadimethoxine equivalents were found in the bile followed by the intestine, liver, blood, skin, kidney, spleen, gill, muscle and fat respectively. Metabolite analysis of select tissues demonstrated that the major metabolite produced was the N-acetylsulfadimethoxine. This metabolite comprised 86% of the total radioactivity in bile and 55% in the liver 20 h after dosing by gavage with sodium ^{35}S-sulfadimethoxine (42 mg/kg). Mass balance studies demonstrated that during the first 25 hours branchial and urinary elimination accounted for 0.6% and 5.5% of the total dose administered respectively. The major constituents for branchial and urinary elimination routes were sulfadimethoxine (76%) and N_4 acetylated sulfadimethoxine (52%) respectively.

Catfish handle sulfadimethoxine in a similar fashion to that found for rainbow trout. Studies by Squibb et al. (32), have demonstrated that following intravenous administration of (40 mg/kg) sodium sulfadimethoxine, plasma elimination also followed a bi-exponential model. The $t_{½ß}$ from plasma was 12.8 h, and the Vd (volume of distribution) 662 ml/kg (*32*). The plasma protein binding was approximately 18% and demonstrated dose independency through a concentration range of 1 - 30 X 10^{-6} M. Oral bioavailability of the sodium salt in catfish was 34.1% with no significant changes upon administration of the free drug. This was in contrast to that reported for trout (*27*). Radiolabel elimination half lives were 12.1, 14.9 and 15.5 h for muscle, liver and skin respectively. Parent sulfadimethoxine constituted greater than 95% of the radioactivity in catfish plasma and muscle. Whereas, greater than 90% of the radiolabel in bile was in the form of N-acetylsulfadimethoxine. Excretion in the urine accounted for approximately 4% of the dose in the first 24 h. Parent (45%), N-acetylsulfadimethoxine (45%) and unidentified polar metabolites (less than 10%) comprised the metabolite profile of urine in catfish (*32*). Elimination routes and metabolite profiles for sulfadimethoxine in catfish appear similar in scope and magnitude to those reported in trout.

Ormetoprim. Ormetoprim is a broad spectrum antimicrobial which is a competitive inhibitor of dihydrofolate reductase. The drug is most often used to potentiate the anti-folate effect of sulfa drugs. The combination of sulfadimethoxine and ormetoprim is commonly employed in the treatment of bacterial outbreaks in aquacultured fish. Plasma concentration time curves following intravenous administration of ormetoprim have been described in catfish triexponentially with α, ß, and γ phases of 0.39, 4.9, and 49 h (*48*). Similar studies in trout have demonstrated $t_{½α}$ and $t_{½ß}$ values of 0.54 and 17.5 h respectively (*31*). Total body clearances (Cl_b) and apparent volumes of distribution (V_{ss}) were similar for both species in the respective independent studies. The large volume of distribution (4,851 ml/kg for trout; 4,966 ml/kg for catfish) suggests ormetroprim is widely distributed and concentrated in tissues.

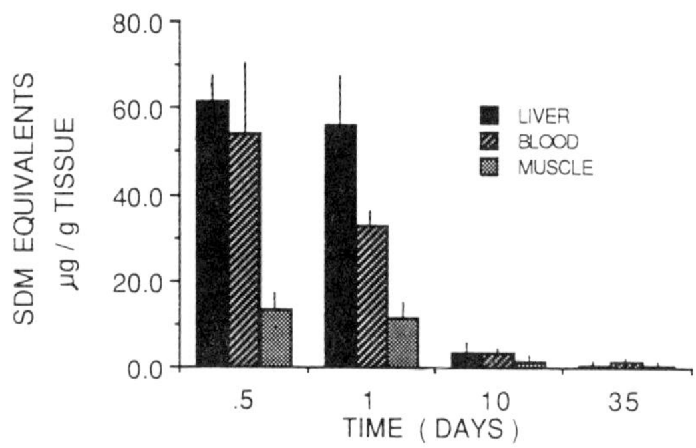

Figure 2. Concentration of total radioactivity in select tissues after oral administration (42 mg/kg) of ^{35}S-sulfadimethoxine (SDM), to rainbow trout (*Oncorhynchus mykiss*) (n = 4 for each time point).

For trout, the peak plasma level following oral administration was 12 h with an apparent bioavailability of 87% (*31*). Bioavailability of ormetoprim was substantially lower for catfish with a value of 52% and peak plasma levels at 6 h (*48*). Tissue distribution studies in trout following oral administration revealed the highest concentration of ^{14}C-ormetoprim in the bile, kidney and liver. Significant ^{14}C-ormetoprim derived residues were evident in the skin (0.90 ppm) and muscle (0.15 ppm) at 38 days. Tissue half-lives of the drug were 99 and 19 days for the skin and muscle respectively (*31*). Residues in trout were more persistent than in catfish. At 72 hours after dosing, concentrations in catfish muscle and skin were 0.04 µg/g and 0.71 µg/g respectively (*48*).

As observed with other drugs, ormetoprim exhibited a low percentage of plasma protein binding in fish. For trout, this averaged 31%. This binding was nonspecific and non-saturable. Plasma protein binding of ormetoprim also has been described as non-specific in other species such as sculpin, skate, shark and flounder (*53*).

Metabolism studies with ^{14}C-ormetoprim in catfish provided evidence for four metabolites in urine (*48*). A total of 21.1% of the intravascularly administered dose was eliminated in the urine within the first 48 h. Of this total 16.4% were polar metabolites, 1.0% was a nonpolar metabolite and 3.7% was the parent drug. In addition, less than 6% of the total dose accumulated in the bile. The authors hypothesized that the balance of elimination occurred by branchial routes, although this was not measured.

Temperature and dietary related effects have been observed in the metabolism and distribution of ormetoprim in catfish (Kleinow, K.M. Louisiana State University, unpublished). Tissue distribution of ormetoprim under the conditions of warm water (28°C) with a semi purified diet supplemented with primarily saturated lipids and coldwater acclimated fish (11°C) with an unsaturated lipid supplementation demonstrated differences between groups in regard to magnitude and time to peak tissue concentrations. In general for most organs the warm water saturated group demonstrated earlier and higher tissue concentrations than cold/unsaturated lipid treated animals. Similarly, differences were observed for hepatic metabolite profiles for the two groups. Throughout the time course a much greater degree of metabolism (30%) occurred for the warm saturated group as compared to the cold unsaturated group. For most time periods a major shift occurred in the production of a single polar metabolite. This metabolite, one of six isolated in the bile, was consistently found in greater amounts in warm saturated groups. Further studies are delineating the temperature and dietary influences of the observed changes.

Oxolinic Acid. Oxolinic acid, a quinoline antibiotic, is widely used in humans for urinary tract infections. This drug has shown efficacy for the treatment of a variety of gram-negative bacterial diseases in fish; however, it has not yet received approval for such use in this country. Recently, studies have investigated the relationship of species differences in metabolism and pharmacokinetics to preferred acclimation temperatures of fish (*54*). These intravenous studies with oxolinic acid in catfish and trout have demonstrated that both species and temperature related differences exist independently. That is to say, species differences are not solely a result of differences in their preferred acclimation temperature. Of the pharmacokinetic data, volume of distribution (V_{ss}), total body clearance, (Cl_b) and $t_{½ß}$, were most dramatically affected (Table III). Lowering catfish acclimation temperature 10°C (to 14°C) resulted in somewhat longer $t_{½ß}$'s (59.9h) than warmwater (24°C) acclimated fish (45.5h). These $t_{½ß}$ values were still significantly shorter than those calculated for trout at 14°C (120h). Interestingly, total body clearance of oxolinic acid from trout at 14°C was more similar to warm acclimated catfish (24°C) than to catfish at 14°C. In catfish the lower acclimation temperature dramatically limited the volume of distribution (Vss) and reduced by 1/3 the total body clearance as compared to warm acclimated animals.

Table III. Pharmacokinetic Values for Oxolinic Acid in Rainbow Trout (14°C), Warm Water Acclimated Catfish (24°C) and Coldwater Acclimated Catfish (14°C) Following a Single Intravenous Dose (5 mg/kg)

Parameter	Trout (14°C)	Catfish (24°C)	Catfish 14°C)
$t_{½ß}$ (h)	120.4 ± 41.2	45.5 ± 9.9	59.9 ± 12.2
V_{ss} (ml/kg)	2168 ± 411	987 ± 202	331 ± 217
Cl_b ml/kg/h	14.2 ± 6.5	15.3 ± 3.1	5.3 ± 2.4

Values represent mean ± SD.
$t_{½ß}$ - terminal elimination phase half-life.
V_{ss} - Volume of distribution
Cl_b - Total body clearance
Adapted From (*54*)

Several studies with rainbow trout have examined the influence of temperature on tissue residue retention of oxolinic acid. These studies have not identified a clear temperature/residue relationship as has been identified with other compounds such as oxytetracycline (*55*) . The range of results reported indicate greater residue persistence with colder temperatures (*56*), no residue-temperature dependency (*57*) and longer residue residence times with warmer temperatures (*58*). It is difficult to extrapolate between studies as dosages and experimental methods varied markedly.

Water quality has been shown to affect oxolinic acid uptake and its pharmacokinetic parameters when fish are exposed via the water. Endo and Onozawa (*59*) reported in studies with Ayu that the volume of distribution of oxolinic acid was dependent on water salinity. In these studies animals exposed to oxolinic acid in freshwater had a volume of distribution 5 times greater as compared to those exposed in 12.5% diluted artificial seawater. The pH has also been shown to influence oxolinic acid uptake from bath exposures (*60*). In this case, uptake across the gill was greater for pHs below the pKa of the drug. This directly follows the ionized/ nonionized species relationship of Henderson Hasselbach. Uptake across the gill occurs primarily in the non-ionized form.

Work by Ishida (*61*) comparing biliary metabolites in seven species of teleosts has demonstrated that for unconjugated metabolites, only parent oxolinic acid was detected in all species. However, glucuronides of oxolinic acid, and of its 7-hydroxy, -6-methoxy and 6- hydroxy -7-methoxy derivatives were found for the rainbow trout, Japanese eel, and tilapia, all freshwater species (Figure 3). Interestingly in the marine species red sea bream, Japanese mackerel, yellowtail and Japanese flounder only the 7-hydroxy - 6-methoxy derivative and 6- hydroxy -7-methoxy derivative glucuronide conjugates were found. Studies using radiolabeled oxolinic acid in rainbow trout have separated chromatographically a total of four metabolites and the parent. All of these compounds were more polar than the parent (*54*).

Tricaine methanesulfonate (MS-222). Tricaine methanesulfonate (3-aminobenzoic acid ethyl ester methanesulfonate; MS-222) is widely used for the sedation and anesthetization of fish. The compound is 0.01% ionized at body pH and has a partition coefficient of 312. This relatively nonpolar lipophilic drug is rapidly absorbed from water and eliminated in freshwater and marine fish species.

Tricaine is metabolized to form a number of hydrolysis products and acetylated amino congeners (Figure 4). In the dogfish shark (*Squalus acanthias*) about 5% of the drug is metabolized by hydrolysis and/or acetylation to 3-aminobenzoic acid and 3-acetylamino benzoic acid. These products are excreted by the kidney. Approximately 95% of the drug is cleared by the gill within 2 hours of administration. The vast majority of branchial elimination is as the parent drug while a small fraction (<10%) exists as the N-acetyl derivative (*62*). Other studies in the rainbow trout have indicated that blood levels of MS-222 parent were approximately 75% of the anesthetic concentration (*63*). Of the drug eliminated by the trout in the urine (15-21%) approximately (77-96%) was acetylated (*63*). It was demonstrated that the blood and urine were cleared of the drug by 8 and 24 hours respectively. In both the dogfish shark and rainbow trout intraperitoneal injections did not induce anesthesia. The elimination by the gill prior to access to the brain is a contributing factor postulated for these results.

Methods for the Study of Drug Disposition in Crustacea

This section will address the characteristics of crustacea that differ markedly

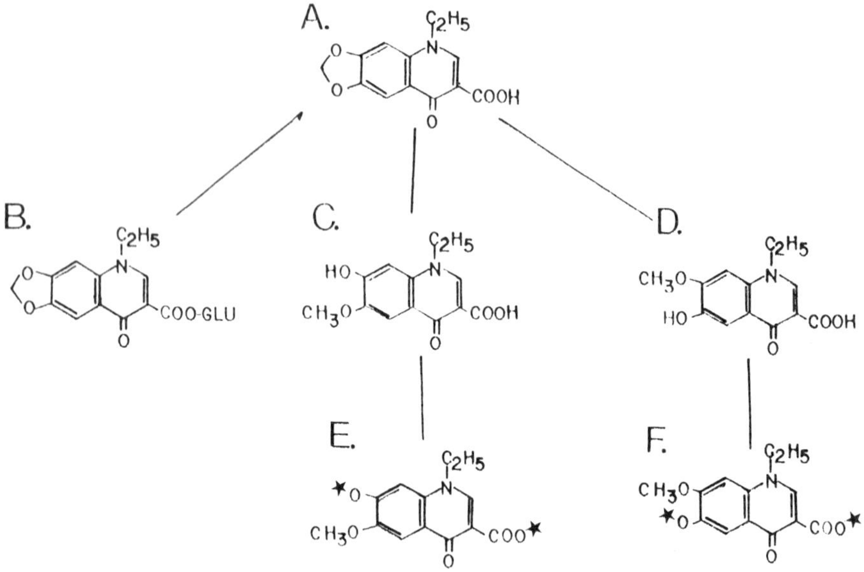

Figure 3. Biotransformation cascade of oxolinic acid in fish. A, oxolinic acid; B., Glucuronide of oxolinic acid; C., 7-hydroxy-6-methoxy derivative (1-ethyl-1,4 dihydro-7-hydroxy-6-methoxy-4-oxo-3-quinoline carboxylic acid) D., 6-hydroxy-7-methoxy derivative (1-ethyl-1,4-dihydro-6-hydroxy-7-methoxy-4-oxo-3-quinoline-carboxylic acid. E,F. Stars represent unconfirmed sites of glucuronidation of 7-hydroxy-6-methoxy and 6-hydroxy-7-methoxy oxolinic acid derivatives.

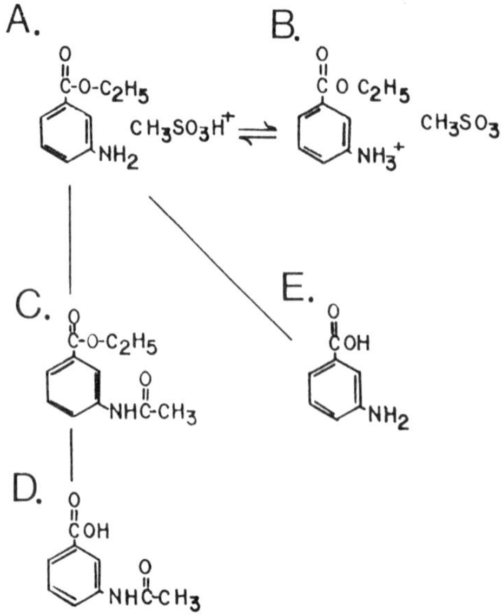

Figure 4. Biotransformation cascade of Tricaine methanesulfonate (MS-222) in fish. A,B., Tricaine methanesulfonate equilibrium; C. Ethyl-3-acetylaminobenzoate; D., 3-acetylaminobenzoic acid; E., 3-aminobenzoic acid.

from other animals, and how these should be taken into account in the design of studies of drug disposition.

Anatomical and Physiological Considerations. It is obvious that crustacean species differ considerably from vertebrates in their biological make-up, and that these important differences should be considered in planning xenobiotic disposition studies.

Circulation and Blood. The circulatory systems of crustaceans are less well developed than those of vertebrates and the total turnover time for hemolymph (blood) in lobsters varies between 1 and 8 minutes (*64*). In lobsters, the hemolymph volume is usually large and has been established at 20 to 28% body weight in the intermoult stage (*64,65*). Thus, initial distribution of a bolus dose into the pericardial sinus (iv) may take longer than in vertebrates. The large hemolymph volume and ready access through hemoceles at each joint makes taking multiple samples of hemolymph from crustacea (especially lobsters) simpler than taking blood from fish. After sampling, crustacean hemolymph clots very rapidly to a solid gel which is not readily disrupted. The clotting process may be slowed or prevented by keeping samples ice-cold, or by adding N-ethylmaleimide, potassium oxalate or magnesium sulfate to hemolymph samples, although in xenobiotic disposition studies, it is advisable to process samples for drug analysis as rapidly as possible. Another notable difference is that the oxygen carrier protein in crustacean's hemolymph is not an iron-containing hemoglobin but a copper-containing hemocyanin (*64*).

Internal Organs. The liver, and to a lesser extent, the kidney, intestine and lung are the most important organs of xenobiotic metabolism in terrestrial vertebrates. Though not functionally identical, analogous organs in crustaceans are the hepatopancreas (sometimes called the digestive gland or mid-gut gland), antennal (or green) gland, intestine and gill, respectively. Figure 5 shows the location of the internal organs. The hepatopancreas is the major organ of metabolism and storage of nutrients, including lipid and glycogen reserves (*66*). In intermoult lobsters, the hepatopancreas is 4 to 5% of body weight and in well-fed lobsters up to 20% of the hepatopancreas weight is fat (James, unpublished observations). The high fat content makes the hepatopancreas a storage site for lipophilic xenobiotics. The antennal glands are paired organs which function in urine formation, salt balance and steroid biosynthesis (*64,67*). Urine forms in two urinary bladders, which empty through small pores at the base at the antennae. The crustacean intestine is much smaller than that of vertebrates, and its role in xenobiotic metabolism has not been investigated. The gills of crustacea, like those of fish, function in oxygen transfer, and may serve as a site of uptake of water-borne xenobiotics. The role of the gill in xenobiotic metabolism and excretion is not as well defined in crustacea as it is in fish, although there are a few examples of gill excretion of xenobiotics (*68*).

Exoskeleton. A major difference between crustacea and fish is the presence of the hard exoskeleton, or shell, which must be shed periodically to allow growth. The shell is a layered structure which contains chitin, protein, lipid, various calcium salts and hydroxides, and pigments, and which appears to adsorb certain xenobiotics (*69*, James unpublished).

Design of Pharmacokinetic and Disposition Studies

General Considerations. The general design of protocols, such as number of animals studied for each data set, hemolymph sampling times and sacrifice times for tissue distribution studies should be similar to those in practice for other species. Drug disposition, and pharmacokinetics of elimination after intravascular administration should generally be studied, even though this will probably not be the actual dosage route, in order to develop comparative data to determine the drug's bioavailability following oral or water-borne routes of administration. Since metabolism and excretion of lipophilic drugs and their metabolites is usually slower in crustaceans than in other species, it may be necessary to hold groups of dosed animals for sacrifice at two to three months after the dose in order to ensure adequate depletion of drug residues to acceptable concentrations in edible tissues. In order to conduct disposition studies accurately , it is best to use radiolabelled drug. The tank water eluent should then be passed over charcoal filters or other suitable materials to remove the radiolabelled drug and metabolites before release of the waste-water.

If a complete mass balance is desired, it is possible to cannulate the urinary papillae (*70*) and the anal orifice (*71*) in some crustacean species by gluing cannulae in place with cyanomethacrylate glue. Cannulated animals are best suited for relatively short term studies (less than 1 week). If excretion across the gills is to be studied, the animal can be held in a static tank with a low water volume for up to 2 hours after dosing, and the tank water saved for analysis (*68*). Alternatively, the tank water can be passed over columns designed to remove the drug and metabolites, and the absorbed material later desorbed and analyzed (*72*). Multiple sequential hemolymph samples for pharmacokinetic analysis are best taken from the hemoceles at the base of each of the walking legs. After the appropriate time, the animal can be anesthetized by chilling in ice, and partially exsanguinated by cardiac puncture. If the animal was not cannulated, terminal urine samples could be collected from each bladder by syringe through the urinary papillae prior to dissection. About 5 to 8 ml of urine can usually be collected from a 500g lobster. The internal organs are relatively fragile, and special care must be taken in removing the parts of the gastrointestinal tract (stomach, hepatopancreas, intestine and intestinal contents) and the antennal glands. Muscle samples should be taken from the center of the tail, well away from the intestine. With the exception of gill and shell samples, which should be analyzed for radioactivity by complete oxidation and scintillation counting, lobster tissues dissolve readily in NaOH and can be prepared for scintillation counting after neutralization of the digest with HCl (*69*).

Routes of Administration

Intravascular Administration. The easiest site of injection is the pericardial sinus, located just above the junction of the abdomen and the tail (Figure 5). Aqueous injection vehicles, pH 6.5 to 9, are well tolerated, as is a small volume (up to 1 ml/kg) of dimethylsulfoxide (DMSO).

Oral Administration. Incorporation of drugs into the food is a major route of actual administration of therapeutic agents. Bioavailability studies may be performed in lobsters by administering the drug by gavage, and subsequently taking hemolymph and tissue samples at various times, as described above. The following protocol has been used successfully in studies with lobsters (*73*). The drug was evenly dispersed in homogenized shrimp of a consistency that could be withdrawn into a syringe. The syringe was fitted with a plastic pipet tip with the end cut off. The lobster was fasted for one day before dosing. The lobster was held on its back and the syringe tip placed between the mandibles into the stomach for delivery of the dose. After removal of the emptied syringe, the lobster was kept on its back for one minute before being returned to the water. The lobster was closely monitored for up to 30 minutes for signs of regurgitation or leakage of the dose. Regurgitation rarely occurred if the syringe was properly placed in a fasted animal.

Water-borne Administration. For water-borne administration, drugs are usually prepared in a bath in which the animals are held for a period of time before introducing fresh untreated water. It is important to ensure that the drug is completely dissolved in the well-mixed water before adding the animal. This method of exposure has not to our knowledge been used to study drug uptake by crustaceans, but has been widely used for studies of the uptake of pesticides and environmental pollutants by crustacea (*74*).

Biotransformation and Identification of Metabolites

Although the separation and identification of metabolites follows the general principles developed by analytical chemists, some of the difficulties encountered in working with crustacea are mentioned.

Isolation from Dosed Crustacea. Metabolites are usually excreted in feces and urine after formation in the hepatopancreas, antennal gland or other sites. The identification of metabolites in crustaceans follows established procedures, i.e., the expected metabolites are predicted from the drug's structure, or from studies with other species, and methods are developed for the extraction and separation of the parent drug and the expected metabolites. Metabolites may then be identified by cochromatography with known metabolite standards, and by various spectroscopic procedures, such as mass and nuclear magnetic resonance. The major difficulty in working with crustaceans lies in isolating metabolites from hepatopancreas, where most of the metabolites of lipophilic drugs are formed. The hepatopancreas generally has a high fat content and also contains detergent-like molecules, such as fatty acyltaurine conjugates, which are thought to mimic the action of bile salts in vertebrates (*75*). The high fat and detergent contents of hepatopancreas interfere with solvent and solid phase extraction procedures, and with chromatographic separation systems. The fats and detergents also hinder the isolation of pure metabolites for positive identification by mass spectrometry and NMR.

***In vitro* methods.** *In vitro* studies with isolated subcellular fractions can provide valuable information on the likely pathways of metabolism and sites of metabolite formation. In conducting studies with hepatopancreas fractions, the possibility of contamination with digestive enzymes and detergents that can inhibit activity must be taken into consideration. In studies with spiny lobster hepatopancreas microsomes, it was necessary to wash the microsomes in a cholate-containing buffer to solubilize inhibitory factors, and to fortify the resulting microsomes-derived, cytochrome P450-containing preparation, with vertebrate cytochrome P450 reductase in order to measure monooxygenase activity (*76*).

Examples of the Disposition of Xenobiotics in Crustacea

The major food producing crustaceans that are presently aquacultured are various shrimp species, crayfish and lobsters (*77*). Pharmacokinetic studies of drugs have been done so far only in the lobster, but studies of the pharmacokinetics and metabolism of some agricultural chemicals have been conducted in crab and crayfish species.

Lobster. Studies have been conducted to determine the pharmacokinetics of excretion, and the oral bioavailability of several drugs in adult lobsters of both sexes. Sulfadimethoxine and ormetoprim, which are used as a drug combination to treat various bacterial diseases, were studied separately and together. The usual dose ratio of the sulfadimethoxine: ormetoprim combination is 42:4 mg/kg. Preliminary studies of the hemolymph concentration, and the tissue distribution at various times after intravascular and oral doses, were conducted with erythromycin, 50 mg/kg, in the free base form. Studies of the fate of phenol were conducted after intravascular administration of several doses.

Sulfadimethoxine. The pharmacokinetics of elimination of the intravascularly administered sodium sulfadimethoxine were independent of dose in the range 21 to 55 mg/kg, and the elimination of half life of parent sulfadimethoxine from hemolymph was 77 hours (*69*). Binding of sulfadimethoxine to hemolymph proteins was dose independent in the range 14 to 203 μg/ml, and was 53% bound (*69*). In the intermoult, gonadally regressed lobsters used in the study, there was no sex difference in drug elimination (*69*). The rate of absorption of orally administered sulfadimethoxine, 42 mg/kg, was more rapid if the water soluble sodium salt were given than if the free drug were given, but in both cases, total bioavailability (area under the curve of hemolymph concentrations of parent drug) was about 50% *(69)*. Figure 6 shows the hemolymph concentration of sulfadimethoxine with time after an intravascular dose of the sodium salt or an oral dose of the free drug. Since the therapeutically used form of the drug is compounded as the non-salt form, the elimination from tissues of drug residues derived from the radiolabelled (^{14}C or ^{35}S) drug was followed after single oral doses of sulfadimethoxine. The tissue distribution of radiolabel at various times after the dose is shown in Figure 7, with % dose shown in part a and tissue concentrations in part b. The highest concentrations of radioactivity were found in the hepatopancreas (Figure 7b) and the intestinal contents (data not shown). Preliminary studies of the composition of ^{14}C in the hepatopancreas showed that, by 2 weeks after the dose, most of the radioactivity was in the form of water-soluble metabolites that were not hydrolysed by glucosidase or sulfatase. N-Acetylsulfadimethoxine was present only in very small quantities, and could have been formed during the extraction and work up. The identity of the other metabolites is presently unknown. It is of interest that concentrations of drug residues in the muscle were the lowest of any of the lobster tissues, including the shell, and were consistently lower than hemolymph concentrations. Muscle concentrations of radioactivity were close to 0.1 mg sulfadimethoxine equivalents/g by 40 d after the dose. By 44 days after the last of multiple doses of sulfadimethoxine (42 mg/kg every other day for 5 doses), muscle concentrations were below 0.1 mg/kg (James and Barron, unpublished data).

Ormetoprim. Studies with intravascularly administered ormetoprim, 4 mg/kg, showed that the drug was cleared from hemolymph and taken up by tissues more rapidly than sulfadimethoxine *(78)*. Parent ormetoprim had an elimination half life from hemolymph of 2 hours (James, unpublished). The hepatopancreas was a major site of uptake of this lipophilic drug following either oral on intravenous administration (Figure 8). The amount of metabolized ormetoprim in the hepatopancreas increased with time, and by one month after the dose most of the radioactivity was in the form of as yet unidentified water-soluble metabolites. As with sulfadimethoxine, mean concentrations of ormetoprim residues (drug-derived radioactivity) were lower in muscle than in hemolymph (Figures 7 and 8), but in the case of ormetoprim, the concentrations of radioactivity in the shell were similar to those of muscle (data not shown).

Erythromycin. Single doses of erythromycin, 50 mg/kg, spiked with macrolide ring-^{14}C-labelled drug, were administered to lobsters by gavage (free base) or by intravascular injection of the phosphate salt. The pH of the injection vehicle was 6.8 to 7.0. After intravascular injection, erythromycin cleared hemolymph with an elimination half life of 16 hours, and the drug reached high concentrations, relative to hemolymph, in hepatopancreas, antennal gland and gill

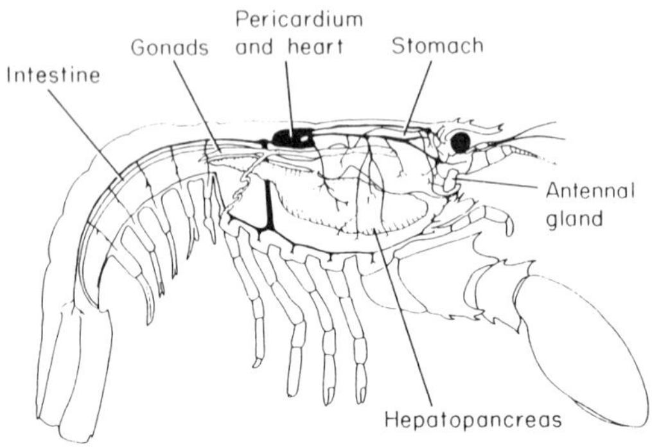

Figure 5. Schematic diagram of the Lobster, *Homarus americanus* showing major organs (adapted from ref. 82).

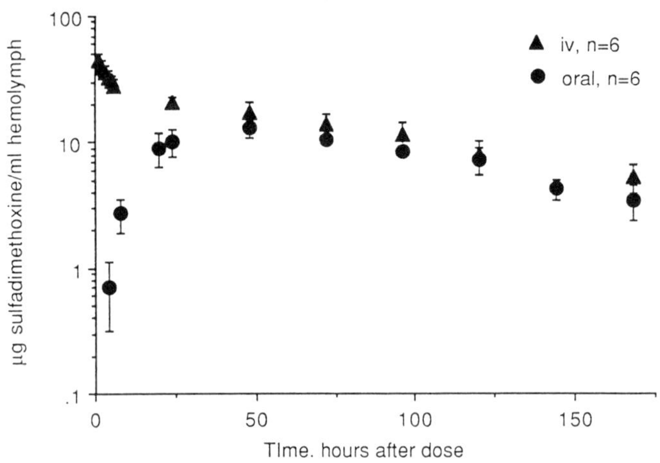

Figure 6. Hemolymph concentrations of sulfadimethoxine after intravascular administration of the sodium salt (*42* mg/kg) or an oral dose of the free drug (*42* mg/kg) to lobsters (*Homarus americanus*).

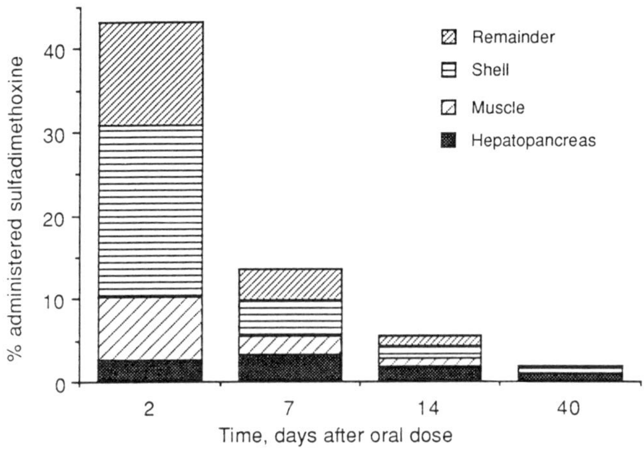

Figure 7a. Tissue distribution of radiolabeled (^{14}C or ^{35}S) -sulfadimethoxine as percent of dose after oral administration (*42* mg/kg) to lobsters (*Homarus americanus*).

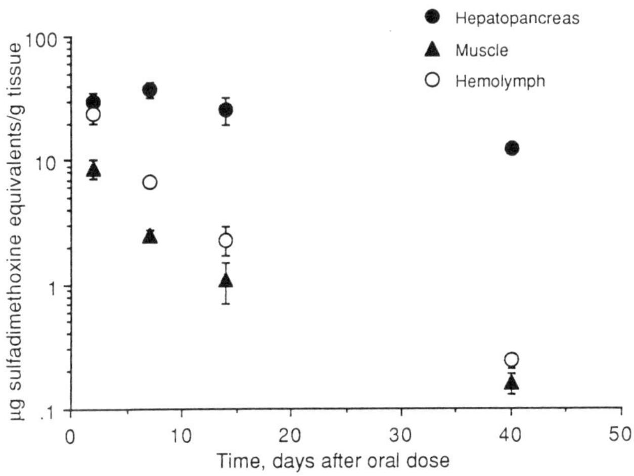

Figure 7b. Concentration of total radioactivity in tissue after oral administration (*42* mg/kg) of ^{14}C- or ^{35}S-sulfadimethoxine to lobsters (*Homarus americanus*) (n = *4-6*).

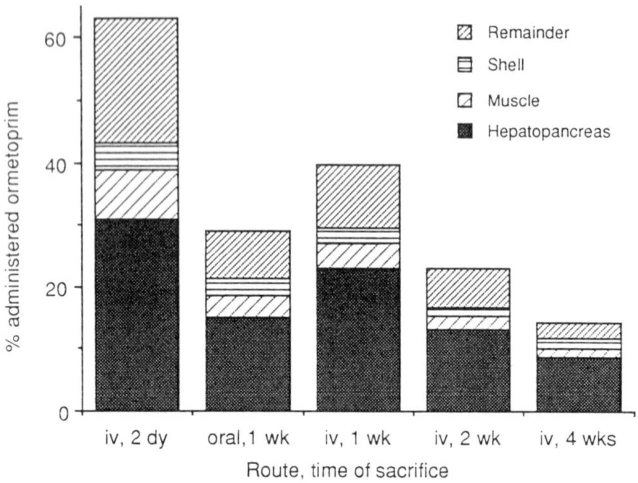

Figure 8a. Tissue distribution of ^{14}C-ormetoprim as percent of dose after either intravenous or oral administration (*4* mg/kg) to lobsters (*Homarus americanus*) (n = *4*)

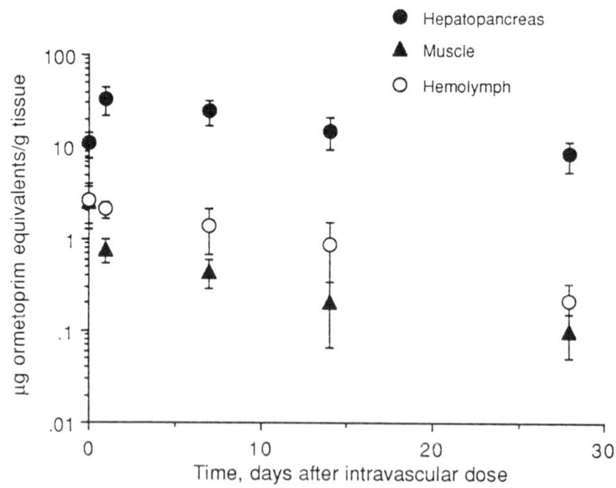

Figure 8b. Concentration of total radioactivity in tissue after intravascular administration (*4* mg/kg) of ^{14}C-ormetoprim to lobsters (*Homarus americanus*) (n = *4*).

(James, unpublished). The organ containing the highest percentage of the dose was hepatopancreas, and erythromycin-derived radioactivity was slowly eliminated in feces (*79*). After oral administration, the hepatopancreas was still the major organ of uptake (Figure 9), but concentrations of radioactivity in antennal gland and gill were considerably lower than after intravascular administration. The radioactivity in hepatopancreas was very slowly converted from parent erythromycin to N-desmethylerythromycin, to small amounts of the anhydro erythromycin breakdown product, and to polar metabolites. By one week after the dose, over 50% of the ^{14}C in hepatopancreas was still parent drug (James et al, unpublished). Erythromycin-derived radioactivity was more slowly eliminated from lobsters than either sulfadimethoxine or ormetoprim, and concentrations of drug residues in muscle were >0.1 mg/kg even at 8 weeks after the dose, the longest time studied (Figure 9). Unlike sulfadimethoxine and ormetoprim, muscle concentrations of erythromycin-derived radioactivity were about 5-fold higher than hemolymph concentrations (Figure 9). These studies indicate that lengthy withdrawal times are needed after administration of erythromycin to lobsters.

Phenol. Intravascularly injected phenol was very rapidly eliminated from hemolymph, with an elimination half life of 14 minutes for parent phenol (*68*). The route of elimination of phenol was dose dependent. At a dose of 0.1 mg/kg, about 35% of the administered drug was eliminated across the gills into the tank water in the first 30 minutes, and the remainder was converted to phenyl sulfate and eliminated from hemolymph into urine with an elimination half life of 6.3 hr in males and 11.9 hours in females (*68*). At doses above 2 mg/kg, >90% of the dose was excreted across the gills. Because its properties are such that it can be either excreted across the gills, or conjugated with sulfate to a readily excreted metabolite, phenol was not retained by the lobster.

Other Decapod Crustacea. The fate of the pesticide fenitrothion has been studied in blue crabs, *Callinectes sapidus*, after water-borne exposure to 5.2 ppb for 48 hours (*80*). The hepatopancreas was the major organ of uptake (0.3 μg fenitrothion equivalents/g after 48 hours), and concentrations in muscle were very low (0.026 μg/g), although muscle concentrations were almost 4 times higher than hemolymph concentrations (*80*). The major metabolites of fenitrothion found in hepatopancreas were fenitrooxon, desmethylfenitrooxon and 3-methyl-4-nitrophenol, and these metabolites were also recovered from the tank water (*80*).

The biotransformation of water-borne methyl parathion has been studied in two prawn species, the Malaysian prawn, *Macrobrachium rosenbergii*, and the ridgeback prawn, *Sicyonia ingentis*, and a crayfish, *Procambarus clarkii*. (*81*). Animals were exposed to ^{14}C-methyl parathion, 0.01 mg/litre for 15 to 24 hours then placed in clean, flow-through water. The tank effluent containing excreted metabolites was passed over XAD-4 resin and the metabolites were subsequently eluted from the resin (*72*). In all three species, the major primary metabolite

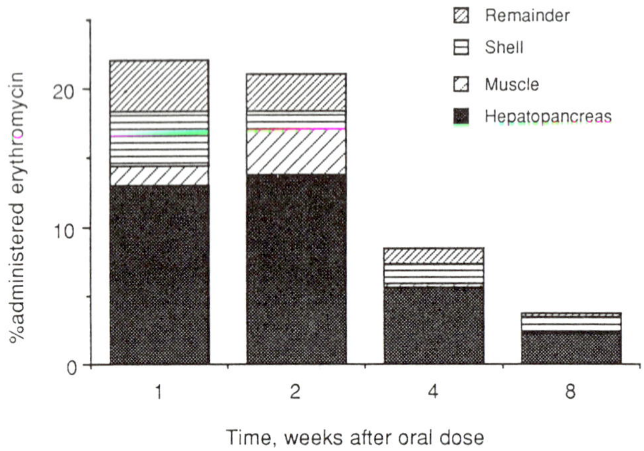

Figure 9a. Tissue distribution of ^{14}C-erythromycin as percent of dose after oral administration (*50* mg/kg) to lobsters (*Homarus americanus*). (n = *6*).

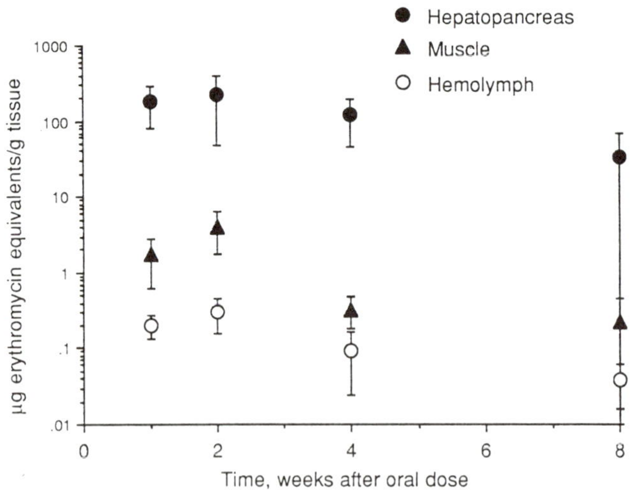

Figure 9b. Concentration of total radioactivity in tissue after oral administration (*50* mg/kg) of ^{14}C-erythromycin to lobsters (*Homarus americanus*) (n = *6*).

was 4-nitrophenol, which was excreted predominantly as sulfate and glucoside conjugates (*81*). These studies did not report tissue concentrations of the pesticide.

The tissue distribution and metabolism of the herbicide triclopyr was studied in crayfish, *Procambarus clarkii*, after static exposure of animals to 1 or 2.5 ppm for 11 days (*25*). After exposure at the higher concentration, tail muscle contained 0.3 µg triclopyr equivalents/g, and the elimination half life was 10 days. The concentration in muscle was more than 10-fold lower than the hemolymph concentration (*25*). The hepatopancreas attained higher concentrations than muscle but lower concentrations than hemolymph at both exposure concentrations. Triclopyr was metabolized in the hepatopancreas to the taurine conjugate and to other polar metabolites (*25*).

The fates of the herbicides 2,4-dichlorophenoxyacetic acid (2,4-D) 2, 4, 5-trichlorophenoxyacetic acid (2,4,5-T) and the DDT metabolite, bis-(4-chlorophenyl) acetic acid (DDA) were studied in the spiny lobster, *Panulirus argus* (*24*). All acids were injected intrapericardially at 10 mg/kg. Both 2,4-D or 2,4,5-T were extensively excreted, unchanged, in urine in the first 24 hours after the dose. About 10% of the dose of 2,4,D and 2,4,5T was taken up by hepatopancreas where the taurine conjugate was formed and subsequently excreted either in urine or feces. Part of the DDA dose was excreted in urine, as unchanged DDA and as DDA-taurine, but DDA was more extensively taken up by shell and hepatopancreas and more slowly excreted than 2,4D or 2,4,5-T. For all three carboxylic acids, muscle concentrations were lower than hemolymph concentrations at 24 hours after the dose, and were less than 5µg/g.

Summary. The available data indicates that compounds that are readily water-soluble, or can be biotransformed into water-soluble conjugates, are more rapidly excreted from crustacea than are lipid soluble drugs that must be metabolized by phase 1 monooxygenases in order to introduce polar functionalities into the molecule. Very lipophilic drugs and pesticides can be expected to attain much higher concentrations in the hepatopancreas than in other tissues, and to be slowly excreted in feces after metabolism to more polar metabolites. For some drugs, hemolymph concentrations are the same order of magnitude as muscle concentrations, but this relationship is different for each drug.

Acknowledgments

The authors would like to acknowledge financial support NIEHS: ESO4184; FDA: FD-U-000158 and FD 01466-01.

Literature Cited

1. Creavan, P.J.; Parke, D.V.; William, R.T. *Biochem. J.* **1965**, *96*, 879-885.

2. Dewaide, J.H.; Henderson, P.T., *Biochem. Pharmacol.* **1968**, *17*, 1901-1907.

3. Buhler, D.R.; Rasmusson, M.E. Comp. *Biochem. Physiol.* **1968**, *25*, 223-239.
4. Pohl, R.J.; Bend, J.R.; Guarino, A.M. *Drug Metab. Dispos.* **1974**, *2*, 545-555.
5. Ahokas, J.T.; Pelkonen, O.; Karki, N.T. *Cancer Res.* **1977**, *37*, 3737-3743.
6. Bend, J.R.; James, M.O. In *Biochemical and Biophysical Perspectives in Marine Biology*; Malins, D.C.; Sargent J.R. eds.; Academic Press: New York, **1978**, Vol. 4, pp. 125-188.
7. Elcombe, C.R.; Lech, J.J. Toxicol. Appl. Pharmacol. **1979**, *49*, 437-450.
8. Stegeman, J.J.; Woodin, B.R. *Fed. Proc* **1980**, *39*, 1752.
9. Statham, C.N.; Szyjka, S.P.; Menhahan, L.A.; Lech, J.J. *Biochem. Pharmacol.* **1977**, *26*, 1395-1400.
10. Miranda, C.L.; Wang, J.L.; Henderson, M.C.; Buhler, D.R. *Biochem. et Biophys. Acta.* **1990**, *1037*, 155-160.
11. Vodicnik, M.J.; Elcombe, C.R.; Lech, J.J. *Toxicol. Appl. Pharmacol.* **1981**, *59*, 364-374.
12. Addison, R.F.; Zinck, M.E., Willis, D.E. *Comp. Biochem. Physiol.* **1977**, *57C*, 39-43.
13. Kleinow, K.M.; Haasch, M.L.; Williams, D.E.; Lech, J.J. *Comp. Biochem. Physiol.* **1990**, *96C*, 259-270.
14. Franklin, R.B.; Elcombe, C.R.; Vodicnik, M.J.; Lech, J.J. In *Microsomes, Drug Oxidations and Chemical Carcinogenesis*; Coon, M.J.; Conney, A.H.; Estabrook, R.W. eds., Academic Press: New York, NY, **1980**, pp. 833-836.
15. Lech, J.J.; Bend, J.R.; *Environ. Health Perspect.* **1980**, *34*, 115-131.
16. Khan, M.A.; Lech, J.J.; Menn, J. *American Chemical Society Symp. Ser.* No. 99. Am. Chem. Soc., Washington, D.C. **1979**.
17. James, M.O., *Xenobiotica*, **1989**, *19*, 1063-1076.
18. Khan, M.A.Q.; Coello, W.; Khan, A.A.; Pinto, H. *Life Sciences* **1972**, *11*, 405-415.
19. Singer, S.C.; March, P.E., Jr.; Gonsoulin, F.; Lee, R.F. *Comp. Biochem. Physiol.* **1980**, *65C*, 129-134.
20. O'Hara, S.C.M.; Neal, A.C.; Corner, E.D.S., Pulsford, A.L. *J. Marine Biol. Assoc.* UK **1985**, *65*, 113-131.
21. James, M.O.: Bowen, E.R., Dansette, P.M.; J.R. *Chem. Biol. Interact.* **1979**, *25*, 321-344.
22. James, M.O.: In *Xenobiotic Metabolism and Disposition.* Kato, R.; Estabrook, R.W.; Cayen, M.N. eds. Taylor and Francis, U.K., USA, **1989**. pp. 283-290.
23. Schell, J.D.; James, M.O. *J. Biochem. Toxicol.* **1989**, *4*, 133-138.
24. James, M.O.: *Drug Metab. Disp.* **1982**, *10*, 516-552.

25. Barron M.G.; Hansen, S.C.; Ball, T. *Drug Metab. Disp.* **1991**, *19*, 163-167.
26. James, M.O.; Barron, M.G. *Vet. Human Toxicol.* **1988**, *30* (Supplement 1), 36-40.
27. Kleinow, K.M.; Lech, J.J.; *Vet. Human Toxicol.* **1988**, *30*, Suppl. 1, 26-30.
28. Endo, T.; Onozawa, M. **1987**, *Nippon Suisan Gak.* *53*, 551-555.
29. McKim, J.M.; Goeden, H.M. *Comp. Biochem. Physiol.* **1982**, *72C*, 65-74.
30. Heming, T.A. *Am. J. Vet. Res.* **1989**, *50*, 93-97.
31. Droy, B.F.; Goodrich, M.S.; Lech, J.J.; Kleinow, K.M. *Xenobiotica.* **1990**, *20*, 147-157.
32. Squibb, K.S.; Michel, C.M.F.; Zelikoff, J.T.; O'Connor, J.M. *Vet. Human Toxicol.* **1988**, *30*, Suppl. 1, 31-35.
33. Aida, K.; Hirose, K.; Yokote, M. Bull. Jap. Soc. Sci. Fish **1973**, *39*, 1,107- 1,115.
34. Heidiger, R.C.; Crawford, S.D. *J. Fish Res. Bd. Can.* **1977**, *34*, 633-638.
35. Kleinow, K.M.; Droy, B.F.; Buhler, D.R.; Williams, D.E. *Toxicol. Appl. Pharm.*, **1990**, *104*, 367-374.
36. Miller, M.R.; Hinton, D.E.; Blair, J.J.; Stegeman, J.J. *Mar. Environ. Res.* **1988**, *24*, 37-39.
37. Statham, C.N.; Melancon, M.J.; Lech, J.J. *Science.* **1976**, *193*, 680-681.
38. Klaassen, C.D.; Plaa, G.L. *J. Appl. Physiol.* **1967**, *22*, 1,151-1,155.
39. Gingerich, W.H.; Weber, L.J.; Larson, R.E. *Comp. Biochem. Physiol.* **1977**, *58C*, 113-120.
40. Kleinow, K.M.; In *Aquatic Toxicology and Risk Assessment*; Mayes, M.A.; Barron M.G., Eds. ASTM STP 1124; American Society for Testing and Materials: Philadelphia, PA, **1991**, Vol. 14; pp 131-138.
41. McKim, J.M., Heath, E.M. *Toxicol. Applied Pharmacol.*, **1983**, *68*, 177-187.
42. Droy, B.F.; Tate, T.; Lech, J.J.; Kleinow, K.M. *Comp. Biochem Physiol.* **1989**, *94C*, 303-307.
43. Beyenbach, K.W.; Kirschner, L.B. *Am J. Physiol.* **1975**, *225*, 389-393.
44. James, M.O.; Pritchard, J.B. *Drug Metabol. Disp.* **1987**, *15*, 665-670.
45. Plakas, S.M.; James, M.O. *Drug Metabol. Disp.* **1990**, *18*, 552-556.
46. Pritchard, J.B.; James, M.O. *J. Pharmacol. Exptl. Therap.* **1979**, *208*, 280-286.
47. Bungay, P.M.; Dedrich, R.L.; Guarino, A.M. J. *Pharmacokinet.* Biopharm. **1976**, *4*, 377-388.
48. Plakas, S.M.; Dickey, E.W.; Barron, M.G.; Guarino, A.M. *Can J. Fish Aq. Sci.* **1990**, *47*, 766-771.
49. Fabacher, D.L.; *Comp. Biochem. Physiol.*, **1982**, *73C*, 277-283.
50. Fabacher, D.L.; *Comp. Biochem. Physiol.*, **1982**, *73C*, 285-288.
51. Hunn, J.B.; Willford, W.A. *Comp. Biochem. Physiol.* **1970**, *33*, 805.
52. Matsubara, T.; Koike, M.; Touchi, A.; Tochino, Y.; Sugeno, K. *Anal. Biochem.* **1976**, *75*, 596-603.

53. Kleinow, K.M.; McElroy, A.E.; *Bull Mt. Desert Isl. Biol. Lab.* **1990**, *29*, 135-136.
54. Jarboe, H.H.; Kleinow, K.M. *Pharmacologist* **1990**, *32*, 149.
55. Salte, R.; Liestol, K. *Acta. Vet. Scand.* **1983**, *24*, 418-430.
56. Ueno, R.; Okumura, M.; Horiguchi, Y.; Kubota, S.S.; *Nippon Suisan Gak.* **1988**, *54*, 485-489.
57. Jacobsen, M.D.; *J. Fish Diseases*, **1989**, *12*, 29-36.
58. Kasuga, Y.; Sugitani, A.; Yamada, F.; Arai, M.; Morikawa, S.; *J. Food Hyg. Soc. Jap.* **1984**, *25*, 512-516.
59. Endo, T.; Onozawa, M. *Nippon Suisan Gak.* **1987**, *53*, 557-562.
60. Endo, T.; Onozawa, M. *Nippon Suisan Gak.***1987**, *53*, 551-555.
61. Ishida, N.; *Nippon Suisan Gak.* **1990**, *56*, 55-59.
62. Stenger, V.G.; Maren, T.H. *Comp. Gen. Pharmacol.*, **1974**, *5*, 23-35.
63. Hunn, J.B.; Schoettger, R.A.; Willford, W.A. *J. Fish. Res. Bd. Canada.* **1968**, *25*, 25-31.
64. Phillips, B.F.; Cobb, J.S.; George, R.W. *In Physiology and Behavior*; Cobb, J.S.; Phillips, B.F. Eds.; The Biology and Management of Lobsters; Academic Press; New York, NY, **1980**, Vol.1, 2-82.
65. Mykles, D.L. *J. Exp. Biol.* **1980**, *84*, 89-101.
66. Conklin, D.E. In *Physiology and Behavior*; Cobb, J.S.; Phillips, B.F. Eds.; The Biology and Management of Lobsters; Academic Press: New York, NY, **1980** Vol.1, 277-300.
67. James, M.O.; Shiverick, K.T.: *Arch Biochem. Biophys.* **1984**, *233*, 1-9.
68. James, M.O.; Schell, J.D.; Barron, M.G.; and Li, C-L. J. *Drug Metab. Disp.* **1991**, *19*, 536-542.
69. Barron, M.G.; Gedutis, C.; James, M.O. *Xenobiotica*, **1988**, *18*, 269-276.
70. Holliday, C.W. *Comp. Biochem. Physiol.* **1977**, *58A*, 119-120.
71. Synder, M.J.; Chang, E.S. *Gen. Comp. Endocrinol.* **1991**, 82, 118-131.
72. Foster, G.D.; Crosby, D.G. *Environ. Toxicol. Chem.* **1986**, *15*, 1059-1071.
73. James, M.O.; Schell, J.D.; Magee, V. *Bull. Mt. Desert Isl. Biol. Lab.*, **1989**, *28*, 119-121.
74. James, M.O. In *Pesticide Chemistry, Advances in International Research, Development and Legislation*; H. Frehse, Ed.; VCH Weinheim, Germany, and New York, USA, **1991**, pp 277-286.
75. Holwerda, D.A.; Vonk, J.K. *Comp. Biochem. Physiol.* 1973, *45B*, 51-58.
76. James, M.O. *Arch. Biochem. Biophys.* **1990**, *282*, 8-17.
77. Friedman, C.S. *Vet. Human. Toxicol.* **1991**, *33* (Supplement 1), 30-33.
78. James, M.O.; Herbert, A.H. *Pharmaceutical Res.* **1988**, 5 S196.
79. James, M.O. *The Toxicologist* **1990**, *10*, 176.
80. Johnston, J.J.; Corbett, M.D. *Toxicol. Appl Pharmacol.* **1986**, *85*, 181-188.
81. Foster, G.D.; Crosby, D.G. *Xenobiotica* **1987**, *17*, 1393-1404.
82. McLaughlin, P.A. In *Internal Anatomy*; Mantel, L. Ed; The Biology of Crustacea; Academic Press: New York, 1983, Vol.5.

RECEIVED March 19, 1992

Metabolism and Disposition

Chapter 9

Pirlimycin in the Dairy Cow

Metabolism and Residue Studies

R. E. Hornish, T. S. Arnold, L. Baczynskyj, S. T. Chester, T. D. Cox, T. F. Flook, R. L. Janose, D. A. Kloosterman, J. M. Nappier, D. R. Reeves, F. S. Yein, and M. J. Zaya

Animal Health Drug Metabolism, Upjohn Laboratories, The Upjohn Company, Kalamazoo, MI 49001

Pirlimycin hydrochloride (I), a lincosaminide antibiotic, is a new therapeutic agent under development for the treatment of mastitis in the dairy cow. Absorption, distribution, metabolism, excretion and residue decline studies of I have been conducted in the dairy cow following intramammary infusion of an aqueous gel formulation of I into all four quarters of the udder via the teat canals. Total milk residues accounted for only 50% of the dose and the milk residue concentration- time course was bi-phasic. Nearly half of the dose was thus absorbed for systemic circulation. Drug residue concentrations in blood were best fit to a two-compartment pharmacokinetic model. Pirlimycin accounted for ≥ 95% of the drug residue in milk and was excreted predominantly as parent compound in the urine and feces. Pirlimycin sulfoxide was the major residue found in the liver, the target tissue for residue analysis. GI tract microflora converted part of the fecal drug residue to 3-(5'-ribonucleotide) adducts of pirlimycin and pirlimycin sulfoxide. The comparative metabolism of I in the rat following oral administration was nearly identical to that in the cow following intramammary infusion.

Pirlimycin (I), Figure 1, is a semi-synthetic member of the lincosaminide antibiotics derived from lincomycin (II) and clindamycin (III). Its activity against most gram-positive organisms is comparable to clindamycin and several times as active against *Staph.aureus* (*1,2*). The proposed use of pirlimycin as a therapeutic agent for the treatment of bovine mastitis was initially investigated by Yancey in an *in vitro* lactating mouse model developed for estimating mastitis activity (*3*). Pirlimycin hydrochloride is now under development as a dairy cow mastitis therapeutic agent. The development of any drug or chemical entity targeted for food-producing animals must undergo a six-step safety evaluation process encompassing the study of the adsorption, distribution, metabolism and excretion (ADME studies) to address drug residue concerns in the consumable products as outlined in the landmark papers by

0097-6156/92/0503-0132$06.00/0

Perez (*4*) and Weber (*5*). This report describes part of the ADME studies carried-out for the safety evaluation of pirlimycin in the dairy cow for the treatment of mastitis.

All four udder quarters of twelve dairy cows in mid-lactation were treated at 4 times the therapeutic dose of 50 mg/ quarter by intramammary infusion of an aqueous gel containing 200 mg of pirlimycin free base equivalents, including the labeled ^{14}C-pirlimycin hydrochloride. The pharmacokinetic parameters for total pirlimycin residue in blood and milk were determined. Three cows were sacrificed at each of four post-treatment intervals (4, 6, 14, and 28 days) to establish tissue residue depletion kinetics. Metabolite profiles of the residues in milk, liver, urine and feces were obtained and the unknown radiolabeled components identified.

A comparative metabolism study in the rat was also conducted to address its relevancy as a toxicological test species. The residue concentration decline kinetics were also determined for pirlimycin in milk from cows treated at a dose rate of 50 mg per quarter in all 4 quarters to establish a milk discard interval for the proposed use.

Materials and Methods

^{14}C-pirlimycin. ^{14}C-Pirlimycin hydrochloride was synthesized by the sequence shown in Figure 2. Final purification was accomplished by recrystallization to a radiochemical purity >98% as measured by HPLC radioactivity monitoring techniques (HPLC/RAM). The specific activity of the purified material was 11.7 mCi/mmole.

Cow animal husbandry for radiolabeled studies. Twelve Holstein cows in mid-2nd or mid-3rd lactation were housed individually in stainless-steel metabolism stalls and maintained therein through 4 or 6 days post-last-treatment then allowed freedom of movement in an enclosed corral until sacrificed. The cage floor was fitted with a plastic-coated rubber mat to reduce the stress of standing for long periods of time. The cages were equipped with manual-fill feed bins and automatic-fill water troughs as well as a drainage system to collect urine and a rear access door to approach the animal for milking and feces collection.

Dose preparation and administration. A mixture of ^{14}C-pirlimycin hydrochloride and non-labeled pirlimycin hydrochloride (99% chemical purity) was prepared to adjust the specific activity to $\leq$ 10,000 dpm/μg. The dose formulation was prepared to contain total pirlimycin free base equivalents at 20 mg per mL in an aqueous gel containing 2% by weight carboxymethylcellulose. Plastets, polyethylene dosing devices used for udder infusions, were filled with 10.1 mL of formulation. The contents of one Plastet was administered into each quarter the udder through the teat canal immediately after milk-out. All four quarters were treated to simulate the rare maximal use situation. A second dose was administered similarly 24 hours following the first dose.

Collection of samples and total residue analysis. Blood (10-15 mL) was collected by jugular venipuncture into heparinized syringes at 0.5, 1, 2, 4, 6, 8, 10, 12, 16, 24, 30, 36, 48, 60, 72 and 96 hours after the 1st dose. Sub-samples of 200-300 mg were immediately weighed in triplicate for radiolabel quantitation by combustion analysis. The remainder of the sample was centrifuged for the separation of plasma, which was

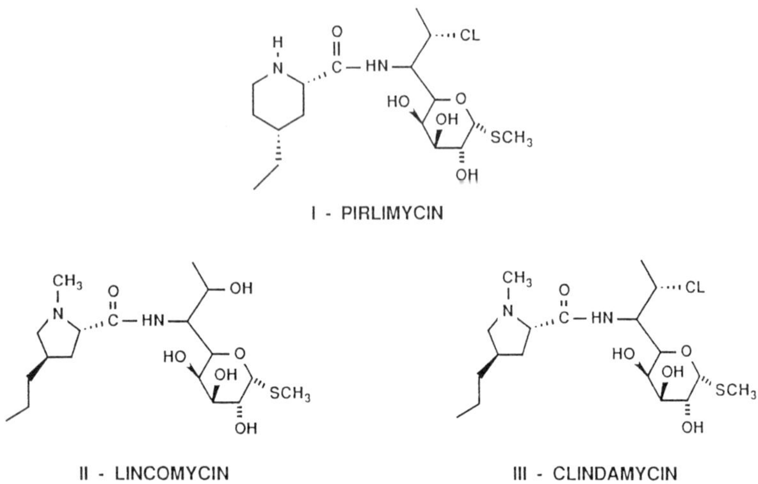

Figure 1. Structures of the lincosaminide antibiotics.

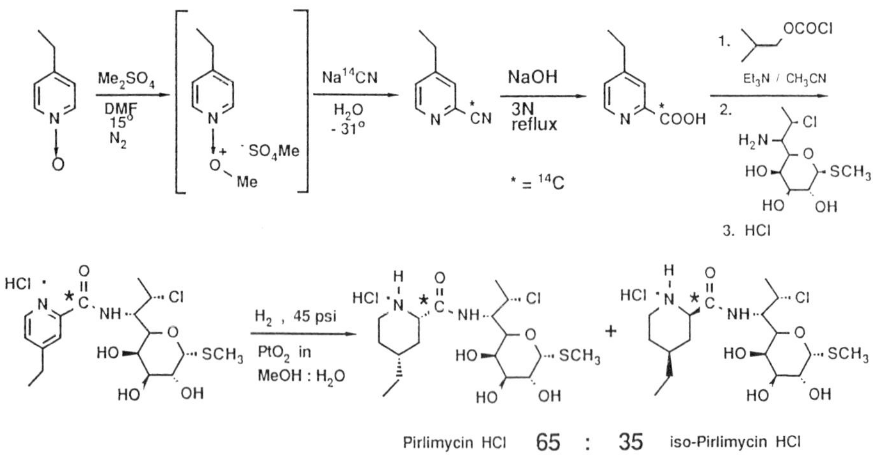

Figure 2. Synthetic sequence for ^{14}C-pirlimycin hydrochloride.

then sub-sampled and stored at -20°C for subsequent analysis. The cows were milked by commercial single cow milking machines at regular 11-13 hour intervals to provide a composite milk sample (4 quarters combined) per cow per time interval. Milk was collected for analysis through 6 days post-last-treatment or until the animal was sacrificed (4 day animals). Urine was collected at 24- hour intervals through the drainage system of the metabolism cage into 5 gallon plastic containers. Each collection was weighed, homogenized and sub-sampled for analysis. Total feces was collected at 24-hour intervals into 5 gallon plastic containers. The net fecal weight was measured and an equal weight of water added to prepare a homogenate slurry for sub-sampling and analysis.

Each animal was sacrificed by captive bolt after the appropriate withdrawal interval (4, 6, 14 and 28 days after the 2nd dose) and processed as in an abattoir. The entire liver, kidneys and udder were excised and 1-2 kg samples of muscle from both the flank and the udder diaphragm and 1-2 kg samples of fat from the abdominal area were collected. Each organ and tissue was minced and processed three times through a commercial meat grinder to prepare respective homogenate samples. Sub-samples (200-300 mg) were prepared in triplicate for total residue analysis. Total radioactivity concentrations, expressed as pirlimycin free base equivalents, were determined by direct liquid scintillation counting (liquids) or combustion analysis (solids) following standard techniques.

Metabolite/residue analysis. **Milk, urine and plasma** samples were first analyzed by a microbiological cylinder/plate procedure against *M.luteus* which has a limit of detection of 0.02 ppm. A sub-sample of the milk was prepared for this assay by a centrifugation step followed by a pH adjustment to 8.5. In addition, an HPLC/RAM analysis was conducted after treating another sub-sample with FTSH (10% formic acid, 30% trifluoroacetic acid, 2% sodium chloride, 2N hydrochloric acid) followed by centrifugation to precipitate the proteins. The supernatant was basified and concentrated by C-18 solid phase extraction (SPE) techniques. The HPLC conditions were: Column - 20 cm x 4.8 mm C-8; Mobile-phase -linear gradient at 5%/minute from 90:10 0.1M pH 7 phosphate buffer:methanol to 20:80; Detectors - UV operated at 214 nm and a radioactivity flow detector operated in the ^{14}C DPM mode.

Tissues and feces were processed as follows: The extraction of >90% of the total ^{14}C residue was accomplished for all samples by homogenizing 1 part sample with 2 parts FTSH, centrifugation, followed by a second extraction of the solids with 20% FTSH. The acid extracts were combined, basified to pH 8.5 ± 0.5 with conc. ammonium hydroxide and processed through C-18 solid phase extraction columns. Pirlimycin and the metabolites were eluted from the columns with methanol and 1% HCl in methanol, respectively. These samples were evaporated to dryness and taken-up in buffer for microbiological or HPLC/RAM analysis.

FAB/MS and NMR. FAB/MS spectra were obtained on a VG magnetic sector instrument. The samples were placed on the FAB target probe containing 2-hydroxy-ethyldisulfide as the matrix solvent. The target was bombarded with xenon at 8-9 KV and the data recorded with a UPACS II data system. Matrix ions were subtracted from the samples ions. All NMR experiments were performed at 25° on a Bruker

AM-500 spectrometer operating at 500.13 MHz for proton magnetic resonance. Samples were prepared in d_6-DMSO and 1-D and 2-D spectra obtained. Several experiments were run for the 2-D spectra: COSY, Relay COSY, HOHAHA, and NOESY which were zero-filled in the F1 dimension only. D_2O exchange spectra were also obtained.

Reference standards. Pirlimycin was obtained as an Upjohn Control Reference Standard of purity >99% as the hydrochloride hydrate. Pirlimycin sulfoxide was prepared from the treatment of pirlimycin with hydrogen peroxide followed by recrystallization. Samples of pirlimycin adenylate and pirlimycin sulfoxide adenylate were prepared by the procedures described by Argoudelis et al (*6*).

Comparative metabolism in the rat. Adult male (6) and female (6) Sprague-Dawley rats were housed individually in polycarbonate metabolism cages and were orally administered by gavage an aqueous solution of ^{14}C-pirlimycin HCl. Five daily doses of 29 mg/kg/day were administered at 24-hour intervals to each rat. Urine and feces were collected at 24-hour intervals just before dose administration. The animals were sacrificed at 2 to 3 hours post-last-dose and liver, kidneys, and samples of flank muscle and abdominal fat carefully excised and placed into tared bottles. Homogenates of 2:1 water:tissue were prepared for combustion/LSC analysis. Metabolite profiles were obtained for liver, urine and feces as described above.

Milk decline study at 50 mg/quarter (1X). Twenty-six lactating cows (Holstein) identified to be mastitic in one or more quarters were treated with two intramammary infusions of 50 mg/quarter of pirlimycin hydrochloride into all four quarters at a 24-hour interval. All cows were milked at 11-13 hour intervals following standard dairy farm practices. Samples from the individual cows (composite of all 4 quarters) were taken for analysis out to 96 hours post-last-dose. The samples were analyzed by the *M.luteus* cylinder/plate microbiological assay.

Results And Discussion

Radiolabeled ^{14}C-pirlimycin was readily synthesized from ^{14}C sodium cyanide and 4-ethyl-pyridine N-oxide as shown in Figure 2. The final reduction step produced a 2:1 mixture of desired product and a biologically inactive stereoisomer. Recrystallization preferentially produced pirlimycin HCL in >98% radiochemical purity and >99% chemical purity with an overall radiochemical yield of 25%.

ADME studies. Twelve cows were administered two doses of ^{14}C-pirlimycin at a dose rate of 200 mg/quarter into all 4 quarters at a 24-hour interval. This dose rate was selected as the highest potential dose rate before the final efficacious dose of 50 mg/quarter had been firmly established. This treatment rate thus resulted in a 4-fold overdose. Blood, milk, urine and feces were collected at various times following the first dose. Combustion analysis of whole blood produced the time course of total residue, as illustrated in Figure 3 for three of the cows. There was a slow absorption of pirlimycin across the udder membrane/blood barrier with maximum concentrations occurring in the 6- to 12-hour posttreatment period. The terminal depletion of the total residue following the second dose appeared to correspond to a two-compartment pharmacokinetic model and suggests a very slow overall elimination. Various pharmacokinetic parameters were derived following noncompartmental analysis (*7*) and are summarized in Table I. Subsequent analysis of plasma, which contained total residue at a concentration approximately equal to whole blood, showed that the plasma residue consisted almost exclusively of unchanged pirlimycin. Thus, the parameters in Table I are useful indicators of the overall pharmacokinetic behavior of pirlimycin in the dairy cow following intramammary administration.

Table I. Whole Blood Pharmacokinetics of Pirlimycin in the Dairy Cow by the Intramammary Route

Parameter	*Mean Value, n = 12*
AUC_{0-120}	2.27 to 7.11 μg/hr/mL
$t_{1/2}$ abs	2.89 ± .46 hours
t_{max} - 1	12 hours
t_{max} - 2	6-12 hours
C_{max} - 1	0.083 ± .03 μg/mL
C_{max} - 2	0.131 ± .047 μg/mL
k_{el}	0.0224 ± .009 hour^{-1}
$t_{1/2}$ el	37.6 ± 17.4 hours

Typical depletion of total residue in milk, expressed as a concentration-time course, is illustrated in Figure 4 for three of the cows. Milk can be treated as an elimination pathway and estimates of pharmacokinetic parameters can be made by the Sigma minus technique (*7*). However, the true focus of milk residue concentration determinations as a function of time is the decline of these residues to levels below the "safe concentration." This will be addressed later when the 50 mg/quarter dose rate study is discussed. The important observation made clear by Figure 4 is the bi-phasic shape of the concentration-time course following the second dose. We interpret this to reflect an initial rapid udder emptying of unabsorbed pirlimycin during the first 2 to 3 milkings post-treatment since each milking is not 100% efficient in milk removal. The slow terminal depletion phase represents systemic elimination of absorbed drug as it is transported back across the udder membrane/blood barrier.

Tissue residues. The concentrations of total residues resulting from the 200 mg/quarter/ dose study in the various tissues at various withdrawal times are presented in Table II. Muscle and fat contained little or no detectable residue beyond day 6. Liver is clearly the target tissue for residue analysis and showed a first order depletion (r = .995) with a $t_{1/2}$ of 5.7 days. Kidney depleted at a faster rate, with a $t_{1/2}$ of 3.3 days. Pirlimycin was not sequestered in the udder as demonstrated by the relatively low concentration of total residue detected in udder.

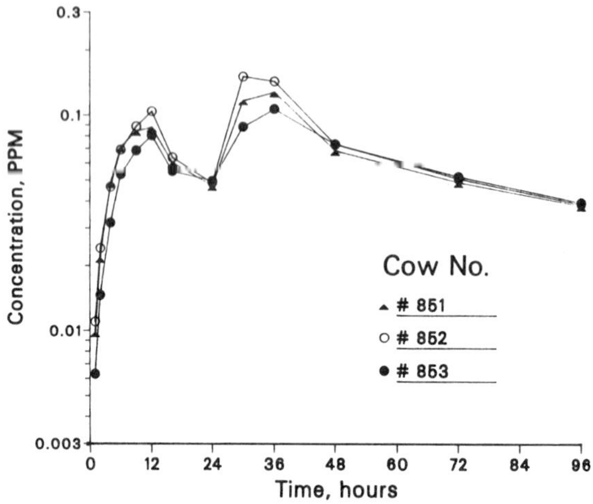

Figure 3. Time-course of ^{14}C-pirlimycin total residue in whole blood in 3 cows treated twice at a 24-hour interval by the intramammary infusion of ^{14}C-pirlimycin hydrochloride into all 4 quarters at 200 mg/quarter.

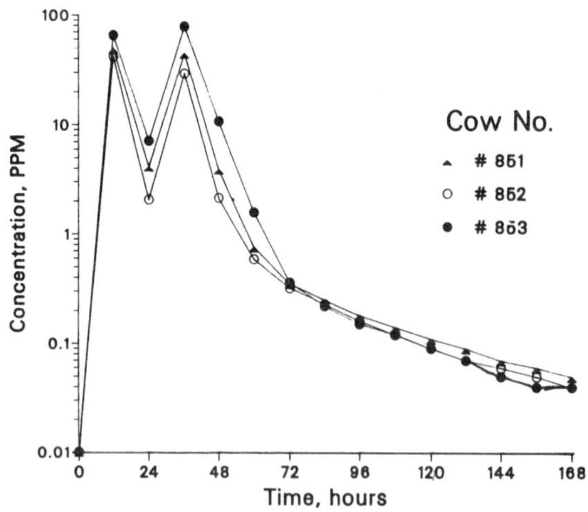

Figure 4. Time-course of ^{14}C-pirlimycin total residue in milk in 3 cows treated twice at a 24-hour interval by the intramammary infusion of ^{14}C-pirlimycin hydrochloride into all 4 quarters at 200 mg/quarter.

Table II. Total ^{14}C Pirlimycin Residues in Tissues - 4X IMM Dose

Tissue	*Mean Concentration (n=3) in Parts Per Million* 4 Day	6 Day	14 Day	28 Day
Liver	9.18 ± 1.37	7.13 ± 1.28	3.57 ± .39	0.50 ± .37
Kidney	1.96 ± 0.71	0.78 ± 0.17	0.26 ± .05	0.01 ± .01
Muscle	0.10 ± 0.04	0.05 ± 0.01	0.02 ± <.01	(0)
Fat	0.22 ± 0.22	0.03 ± 0.01	0.01 ± <.01	(0)
Udder	0.97 ± 0.62	0.13 (n=1)	0.03 ± <.01	(0)

The overall dose accountability is presented in Table III. Note that the total residue in milk accounted for approximately half of the overall dose which suggested that half of the dose was absorbed, i.e. transported across the udder membrane/blood barrier for systemic circulation. Nearly 10% of the total dose was renally excreted and 24% was excreted via the GI Tract through 4 to 6 days of collection. The overall recovery of the administered dose was 89%.

Table III. Disposition and Accountability of ^{14}C-Pirlimycin in the Dairy Cow Following Intramammary Administration

Withdrawal Time, days	*Mean Percent of Total Administered Dose (n = 3)*[1] *Milk*	*Urine*	*Feces*	*Tissues*[2]	*Total*
4	51.6	7.6	22.8	8.9	90.9
6	58.7	10.4	18.3	5.8	91.2
14	42.3	9.4	30.2	2.4	84.3
28	50.9	12.2	23.8	0.3	87.2
MEAN:	50.9 ± 9.8	9.9 ± 2.7	23.8 ± 6.9	(4.4)[3]	88.9 ± 4.9

[1] 200 mg/quarter x 4 quarters/dose x 2 doses, 3 cows per time point.
[2] Calculated from weight of whole liver, kidneys, udder and estimated muscle and fat weights as 55% and 25%, respectively, of total body weight at slaughter.
[3] Mean residue concentration over the withdrawal time range in tissues is for computation only and has no physiological significance.

Metabolite Profiles in Milk. Milk was analyzed by both a microbiological procedure based on the antibiotic concentration-dependent inhibition of growth of the organism *M.luteus* in a standard cylinder/plate assay and by HPLC where the relative percent of pirlimycin was determined by radioactive monitoring techniques. These results, summarized in Table IV, demonstrated that pirlimycin was greater than 95% of the residue, as measured by both procedures, even during the elimination phase when absorbed drug was excreted back through the udder. This suggested that either

Table IV. Concentration and Composition of Pirlimycin Residue in Milk - Dose at 200 mg/quarter, Mean values (n = 12)

*Sample Time**	*Total Residue Concentration by ^{14}C LSC*	*M.luteus Microbio. Analysis*	*Percent Bioactive*	*Percent Pirlimycin by LC/RAM*
D1 + 12 hours	50.64 ppm	48.57 ppm	96	95
D1 + 24 "	5.28 "	5.01 "	95	96
D2 + 12 "	43.95 "	41.48 "	94	96
D2 + 24 "	5.14 "	4.68 "	91	92
D2 + 36 "	1.53 "	1.50 "	98	91
D2 + 48 "	0.79 "	0.78 "	99	99
D2 + 60 "	0.39 "	0.39 "	100	97
D2 + 72 "	0.23 "	0.25 "	109	97
D2 + 84 "	0.17 "	0.19 "	112	95
D2 + 96 "	0.14 "	0.15 "	107	98
D2 + 108 "	0.11 "	0.11 "	100	100
D2 + 120 "	0.09 "	0.09 "	100	86

* D1 = Dose 1, D2 = Dose 2

the metabolism of pirlimycin was minimal or that the metabolites were not circulated through the vascular system (and thus underwent biliary excretion without re-absorption) or transported across the udder-membrane/barrier. Although the data on the plasma metabolite profiles were not exhaustively examined due to the low concentration of residue present, the data indicated that most of the plasma residue was pirlimycin itself. This is consistent with the absence of detectable metabolites of pirlimycin in milk during the terminal elimination phase.

Metabolite Profiles in Liver. The total radioactive residue in all 12 livers was efficiently extracted with a mixture of formic, trifluoroacetic, and hydrochloric acids. After neutralization of the acids with ammonium hydroxide and partial purification by solid phase extraction techniques, HPLC/RAM analysis produced essentially a two-component profile. The minor component was pirlimycin itself in relative amounts of 27.6% ± 9.7%. The bulk of the residue, 77%, was a single substance and was identified by co-chromatography with an authentic standard and by FAB/MS as pirlimycin sulfoxide (PS). The microbiological analysis of the extracts confirmed this relative ratio since PS has an antibiotic activity 142 times less than that of pirlimycin against *M.luteus*. The ratio of pirlimycin to total residue was therefore established and will provide the necessary correlation of residue depletion needed for the calculation of residue marker concentration (Rm) and the establishment of a tissue withdrawal period.

Metabolite Profiles in Urine. Urinary metabolites were profiled by HPLC/RAM techniques and was found to be comprised of pirlimycin (80%), PS (8%), and two minor components (11%), later found to be major components in feces. The presence of these fecal metabolites in urine was apparently due to the method of collection of the urine and feces which allowed the two matrices to co-mingle in the collection trays and thus not true urinary metabolites as will be discussed below.

Metabolite Profiles in Feces. Fecal metabolites were extracted and profiled by HPLC/RAM following the procedures described for liver. Pirlimycin comprised 45% of the overall total residue and PS only 1.5%. Two other components comprised 32% (M1) and 18% (M2), respectively. (The 18% component appeared to potentially be a mixture of two substances, M2 and M3). Partial purification by column chromatographic techniques provided a mixture of substances whose FAB mass spectrum was as shown in Figure 5. The relatively high mass ions at m/z 717, 740, and 762 were suggestive of ribonucleotide adducts reported for lincomycin and clindamycin (*8-11*). The adducts pirlimycin-3(5'-adenylate), m/z 740, and pirlimycin-3(5'-uridylate), m/z 717, Figure 6, were confirmed by comparison to authentic samples prepared following published procedures (*6*) and were assigned as metabolites M1 and M2. The ion at m/z 762 was the sodium adduct of pirlimycin adenylate. The ion at m/z 778 was initially proposed as the sodium adduct of either pirlimycin-guanylate or PS-adenylate.

The results from several 2D NMR experiments indicated that the site of conjugation was at the 3-OH of pirlimycin. Whereas Argoudelis, et al, has reported that the site of conjugation for both lincomycin and clindamycin was at 3-OH (*6*), Brisson-Noel et al provided evidence that clindamycin was conjugated at the 4-OH (*11*). Table V lists the chemical shifts observed for the various sugar ring protons for

Table V. Proton NMR Chemical Shift Data for Pirlimycin Compounds

	Chemical Shift in PPM Relative to d_5-DMSO at 2.49 PPM			
Proton	*Pirlimycin*	*Pirlimycin Sulfoxide*	*P-Adenylate Metabolite M1*	*PS-Adenylate Metabolite M3*
S-CH_3	2.04	2.75	2.02	2.72
C-1 H	5.18	4.66	5.24	4.62
C-2 H	3.90	4.18	4.10	4.38
C-3 H	3.37	3.96	3.92	4.50
C-4 H	3.64	3.79	3.79	3.89
C-5 H	3.97	3.87	4.02	3.82

pirlimycin, PS and the adenylated metabolites. The chemical shifts of the methyl protons indicate that metabolite M3 was the PS-adenylate and not pirlimycin-guanylate. The C-3 proton is shifted downfield by 0.55 ppm for the metabolites compared to the unconjugated standards. The protons at C-2 and C-4 are only slightly shifted downfield (0.1 to 0.2 ppm) relative to the unconjugated species. In addition, the FAB/MS and NMR spectra of these metabolites were confirmed by comparison

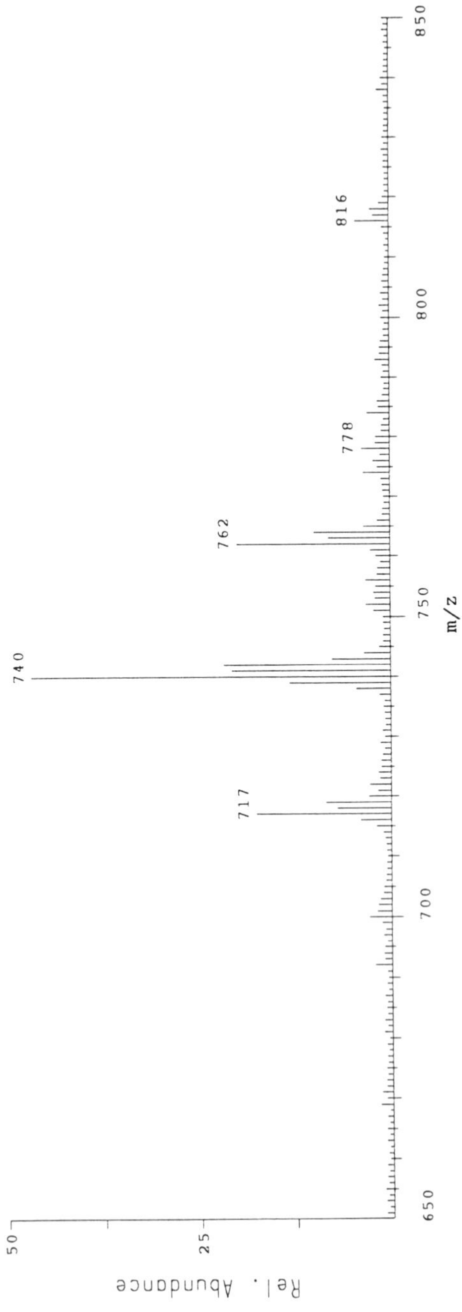

Figure 5. FAB mass spectrum of a mixture of metabolites isolated from the feces of a dairy cow treated twice at a 24-hour interval by the intramammary infusion of ^{14}C-pirlimycin hydrochloride into all 4 quarters at 200 mg/quarter.

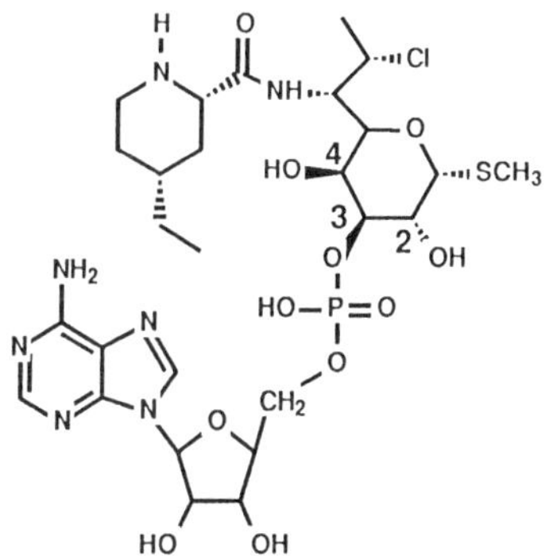

FAB m/z 740
Pirlimycin 3 - (5' - Adenylate)

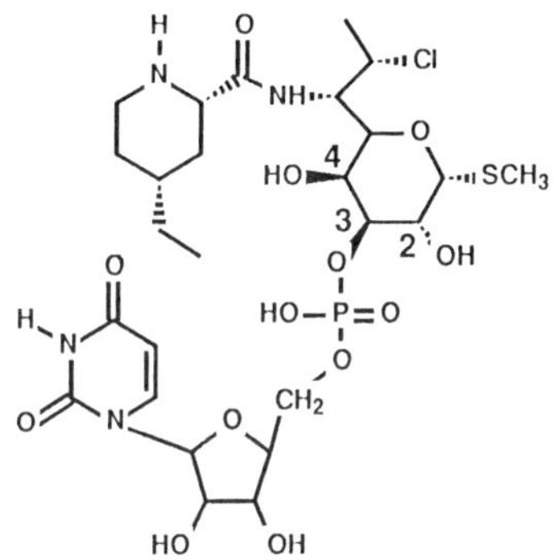

FAB m/z 717
Pirlimycin 3 - (5' - Uridylate)

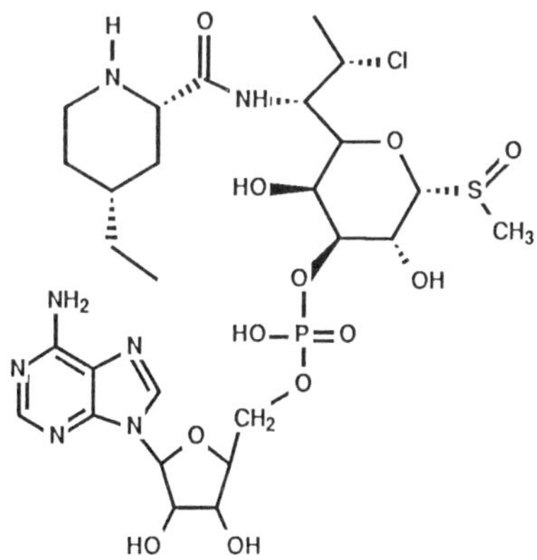

FAB m/z 756
Pirlimycin sulfoxide 3 - (5' - Adenylate)

Figure 6. Structures of the ribonucleotide adducts isolated from the feces of a dairy cow treated twice at a 24-hour interval by the intramammary infusion of ^{14}C-pirlimycin hydrochloride into all 4 quarters at 200 mg/quarter.

to authentic standards, demonstrating that conjugation of pirlimycin to the ribonucleotides was on the C-3 hydroxyl as illustrated in Figure 6.

The generation of these ribonucleotide adducts are believed to be the result of GI Tract microfloral activity and not enzymatic/metabolic transformations in the cow itself. Such adducts have been well documented as products of antibiotic inactivation produced by various strains of *Streptomyces* and *Staphylococci* for a variety of substances such as the lincosaminides (*8-11*), spectinomycin (*12*) and streptomycin (*13*). Furthermore, we have been able to generate pirlimycin 3-(5'-adenylate) by simply incubating pirlimycin with fresh cow manure from untreated cows, although the organisms involved have not been characterized.

Metabolism Scheme. The overall metabolism scheme for pirlimycin in the dairy cow following intramammary administration is summarized in Figure 7. The large majority (68%) of the IMM dose is excreted in the milk, urine, and feces as unchanged parent pirlimycin. A small percentage (4%) appears as pirlimycin sulfoxide generated by hepatic oxidation and is excreted in both urine and feces. Both pirlimycin and PS are partially converted to ribonucleotide adducts by GI Tract microflora and excreted in feces. These polar adducts are apparently not reabsorbed since there was no evidence of their presence in either milk or tissue. There was no evidence of amide bond cleavage nor loss of methyl sulfide, although these metabolic (hydrolytic) transformations could not be absolutely ruled out as minor pathways since 4% of the recovered dose was unidentified.

Comparative Metabolism - Rat. Rats were treated orally with a single daily dose of ^{14}C-pirlimycin at 29 mg/kg/day for 5 consecutive days. This dose rate was the lowest concentration for which a toxicological response was noted in a 90-day chronic study (Jackson, T.; The Upjohn Company, unpublished data. The No Observable Effect Level, NOEL, was 10 mg/kg/day which yields a safe concentration of total residue in muscle, liver, and milk equal to 1.2 ppm, 2.4 ppm, and 0.4 ppm, respectively). Urine and feces were collected at 24-hour intervals throughout the period. The animals were sacrificed at 2 to 3 hours post-last-treatment and liver, kidney, muscle and fat were collected for analysis. Some of the study parameters are summarized in Table VI.

Table VI. Comparative Metabolism of Pirlimycin in the Rat (Oral) and Cow (IMM) - Dose/Residue Data

	Female Rat	*Male Rat*	*Dairy Cow*
No. of animals	6	6	12
Dose Route	Oral	Oral	IMM
Dose Rate	29 mg/kg	29 mg/kg	1600 mg/dose
No. doses	1/day x 5	1/day x 5	1/day x 2
Sac. Interval	2 hours	2 hours	4 to 28 days
Liver Res. Conc.	18.9 ppm	18.8 ppm	9.2 to <1 ppm

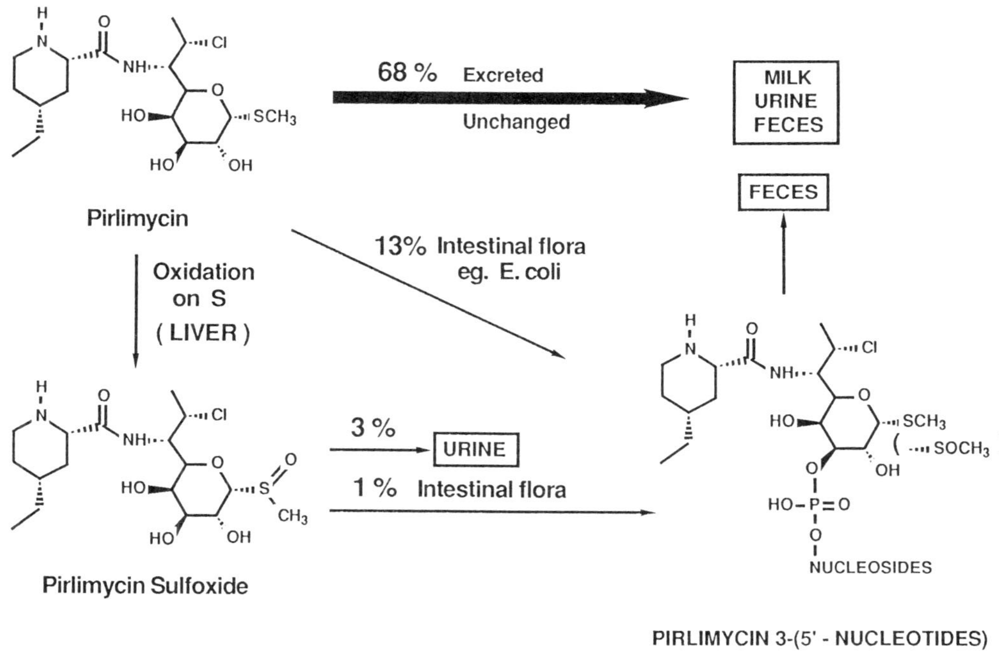

Figure 7. The metabolism scheme for pirlimycin in the dairy cow following intramammary administration of pirlimycin hydrochloride.

Metabolite profiles for urine, feces and liver were established as shown in Table VII. These results showed that pirlimycin was singularly metabolized in the rat liver to the sulfoxide as was observed for the cow. The excreted urinary and fecal residues were again largely unchanged pirlimycin with lesser relative amounts of the sulfoxide. There were a few minor unknowns observed with poor signal/noise values in the HPLC/RAM chromatograms, none of which were isolated in sufficient quantities for identification. They did not, however, appear to be ribonucleotide adducts by virtue of the relative HPLC retention times. Similarly, clindamycin had been observed to be metabolized in the rat to the sulfoxide, although N-demethylation was a route of metabolism observed as well (*14*), a route which is unavailable for pirlimycin. Therefore, these results demonstrated that the rat was a suitable toxicological test species for the safety evaluation of pirlimycin residues generated by the target species.

Table VII. Comparative Metabolism of Pirlimycin in the Rat (Oral) and Cow (IMM) - Metabolite Composition as a Percent of Sample Residue

Sample	*Species*	*Pirlimycin*	*Pirli.Sulfoxide*	*Other*[1]
LIVER	Rat[2]	67	31	2 (u)
	Cow	22	77	1 (u)
URINE	Rat[2]	87	7	6 (u)
	Cow	81	8	11 (RNAs)
FECES	Rat[2]	90	8	2 (u)
	Cow	45	2	53 (RNAs)

[1] u = unknown, RNAs = RiboNucleotide Adducts
[2] average of male and female

Milk Residue Decline Study at 50 mg/quarter. Twenty-six dairy cattle in mid-lactation and identified as mastitic in one or more quarters were given two intramammary infusions of pirlimycin HCl into all 4 quarters of the udder at a 24-hour interval at a dose rate of 50 mg/quarter (1X). Each cow was milked at 11-13 hour intervals and sub-samples taken for microbiological assay. The results are summarized in Table VIII. As observed in previous studies, the decline of the concentration of pirlimycin residue appears to be bi-phasic with a rapid initial depletion followed by a slower terminal elimination phase. Statistical analysis of the residue decline to a concentration below the calculated safe concentration of 0.4 ppm [following FDA guidelines of applying a confidence interval of 95% on the 99th percentile (*15*)] support a 36-hour milk discard interval (48-hour safe milk) for pirlimycin in the US.

Table VIII. Pirlimycin Residue in MILK at 1X Dose (50 mg/quarter x 4 quarters/day x 2 days)

*Sample Time**	*Mean Conc. (26 cows)*	*Standard Deviation*
D2 + 12 hours	8.45 ppm	9.02 ppm
D2 + 24 "	0.87 "	0.80 "
D2 + 36 "	0.23 "	0.18 "
D2 + 48 "	0.11 "	0.06 "
D2 + 60 "	0.07 "	0.04 "
D2 + 72 "	0.05 "	0.03 "
D2 + 84 "	0.03 "	0.03 "
D2 + 96 "	0.02 "	0.02 "

* D2 = Dose 2

Summary

A complete characterization of the absorption, distribution, metabolism and excretion of pirlimycin in the dairy cow following intramammary administration has demonstrated that pirlimycin is readily absorbed from the udder and excreted largely unchanged in milk and urine. Oxidation of the sulfide to the sulfoxide was the only hepatic metabolism observed. However, microflora in the GI Tract of the cow converted both pirlimycin and pirlimycin sulfoxide to ribonucleotide adducts with the addition on the C-3 hydroxyl of the sugar ring. Elimination kinetics of the pirlimycin residue in milk led to a proposed milk discard time of 36 hours for the US market.

Literature Cited

1. Birkenmeyer, R.D.; Kroll, S.J.; Lewis, C.; Stern, K.F., Zurenko, G.E. *J. Medicinal Chemistry*, **1984**, *27*, 216-23.
2. Garcia-Rodriguiez, J.A.; Garcia-Sanchez, J.E.; Prieto, J.; Sanchez de S. Lorenzo, A.; *Antimicrob. Agents and Chemo.*, **1982**, *22*, 893-4.
3. Yancey, R.J.; Kinney, M.L.; Ford, C.W. *J. Antimic. Chemo.*, **1985**, *15*, 219-232.
4. Perez, M.K.; *J. Tox. and Environ. Health*, **1977**, *3*, 837-857.
5. Weber, N.E.; *J. Animal Science*, **1983**, *56*, 244-251.
6. Argoudelis, A.D.; Coats, J.H.; Mizsak, S.A. *J. Antibiotics*, **1977**, *30*, 474-487.
7. Gibaldi, M.; Perrier, D. Pharmacokinetics; Drugs and the Pharmaceutical Sciences Volume 1; Marcel Dekker, Inc.: New York, NY, 1982.
8. Argoudelis, A.D.; Coats, J.H. *J. Amer. Chem. Soc.*, **1971**, *93*, 534-5.
9. Hornish, R.E.; Gosline, R.E.; Nappier, J.M. *Drug Metab. Rev.*, **1987**, *18*, 177-214.
10. Marshall, V.P.; Patt, T.E.; Argoudelis, A.D. *J. Ind. Microbiol.*, **1986**, *1*, 17-21.
11. Brisson-Noel, A.; Delrieu, P.; Samain, D.; Courvalin, P. *J. Biol. Chem.*, **1988**, *263*, 15880-15887.
12. Yamada, T.; Tipper, D.; Davies, J. *Nature*, **1968**, *219*, 288.
13. Benveniste, R.; Yamada, T.; Davies, *J. Infect. Immun.*, **1970**, *1*,109-119.
14. Sun, F.F. *J. Pharm. Sciences*, **1973**, *62*, 1657-1662.
15. Food and Drug Administration "General Principles for Evaluating the Safety of Compounds Used in Food-Producing Animals"; September, 1986.

RECEIVED April 14, 1992

Chapter 10

Albendazole in Cattle Administered via a Sustained-Release "Captec" Device

Pharmacokinetics and Tissue Residues

M. Y. K. Ho, D. W. Gottschall, and R. Wang

Department of Drug Metabolism and Developmental Pharmacokinetics, SmithKline Beecham Animal Health, 1600 Paoli Pike, West Chester, PA 19380

The tissue residue levels and plasma pharmacokinetics of albendazole in cattle administered via a new sustained release 'Captec' device were studied. The daily dosage level released by the device of 0.9 mg/kg was maintained for up to 85 days for the eighteen devices studied. Neither albendazole (ABZ) nor albendazole sulfoxide (ABZ-SO) was detected in the plasma during the entire study period. The plasma albendazole sulfone (ABZ-SO_2) level was 100 fold less than that observed when ABZ is given at the normal commercial single dose of 10 mg/kg b.w. and its presence can be used to monitor the proper functioning of the device. The albendazole marker (albendazole 2-aminosulfone, ABZ-NH_2) residue levels in the liver of 'Captec' cattle at 0- and 5-Day withdrawal were 651 ± 38 ppb and 317 ± 43 ppb, respectively, the majority of which was bound residue. These values are significantly lower than the levels observed at similar withdrawal times when ABZ is given as a single 10 mg/kg dose. Thus, this type of sustained release formulation of albendazole lowered both the plasma and tissue residue levels of albendazole in cattle. Although not proven in this study, it is also highly likely that the relationship between the marker and total residue (Rm) is different from that of the single dose case. This difference may present a regulatory dilemma for the drug sponsor since a uniform method and marker threshold must be applied for each individual drug regardless of the type of formulation.

Albendazole (ABZ) is one of a class of benzimidazoles used to control helminth infections in cattle. It is generally administered as a single oral dose of 10 mg/kg b.w. either by bolus or drench. The residue levels and plasma pharmacokinetic profiles resulting from a single dose of ABZ are well documented (*1*). Recent studies have made important progress into alternative methods of anthelmintic

0097–6156/92/0503–0148$06.00/0

delivery systems. It was found that efficacy can be attained if the anthelmintics are administered at a significantly lower dose but for longer periods of time. In other words, the achievement of a minimum level of parent drug and/or active metabolite(s) in plasma represents the important factor for effective anthelmintic activity rather than a high C_{max} or AUC (2,3). As a result of these findings, devices for the continuous slow release of drug are becoming available. The advantages of these devices are numerous (*4*) and include

(a) convenience in drug administration,
(b) economy and cost savings due to the administration of fewer doses,
(c) reduced handling stress to the animals,
(d) pasture-clean resulting in reduced reinfection rates, and
(e) lower milk and tissue residue levels.

The 'Captec' device (manufactured by 'Captec' Ltd. in New Zealand) is one of the sustained release intra-ruminal devices currently under development for use in cattle. Its design is rather simple (Figure 1) and basically consists of a polypropylene barrel with a plunger and spring on one closed end. The spring is compressed and the barrel filled with seven tablets containing ABZ. Drug is extruded when contact is made between rumen fluid and the ABZ tablet at the orifice on the opposite end of the barrel. The primary factors governing drug release rate include the diameter of the orifice, spring strength and chemical properties of the gel which forms between the rumen fluid and the tablets at the orifice. The 'Captec' device includes a set of "wings" which are secured to the barrel with water-soluble tape prior to administration. Upon contact with the rumen, the tape dissolves, and the wings expand to a predetermined angle, preventing regurgitation of the device.

The metabolism of albendazole has been well studied (*1,5*). The parent compound is short-lived with the major metabolites being generated by a combination of oxidative and hydrolytic processes (Figure 2). One of the major metabolites, albendazole 2-aminosulfone, has been designated as the marker metabolite for which an approved regulatory method has been established in liver, the target tissue.

In the present study, three groups of six cattle (a total of eighteen) were administered the 'Captec' device intra-ruminally via a fistula. Drug release rates were determined by HPLC assay of ABZ and its major metabolites in plasma as well as by physical measurement of the plunger movement. ABZ marker residue levels were determined using the established regulatory procedure in total liver, an ethyl acetate extractable fraction as well as the remaining intractable (bound) residue on each of three cattle at 0- and 5-Day withdrawal.

Experimental

Materials. The albendazole tablets used in Group I and II were formulated by Captec Ltd. and Fernz Corp., respectively, while the Group III tablets were pressed in-house at SmithKline Beecham Animal Health (SBAH). The 'Captec' devices were provided by Captec Ltd. Nylon strings were attached to the devices to facilitate removal from the rumen for the measurement of plunger movement. Albendazole (SKF 62979), albendazole sulfoxide (SKF 77664), albendazole sulfone (SKF 63896), albendazole 2-aminosulfone (SKF 81038), and 5-(n-butylsulfonyl)-1H-benzimidazolyl-2-amine (internal standard, SKF 101437) were obtained in-house.

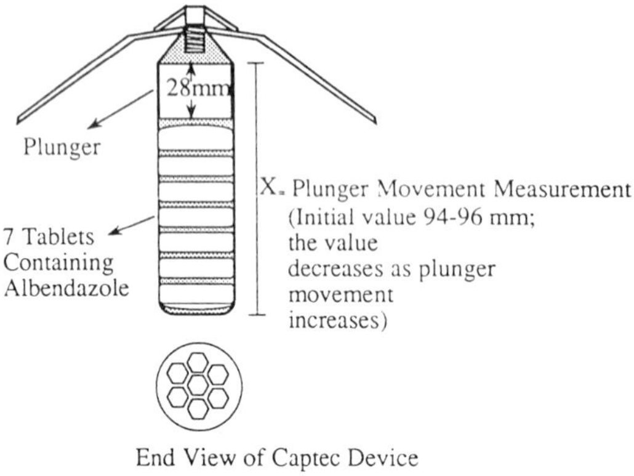

Figure 1 Diagram of A 'Captec' device

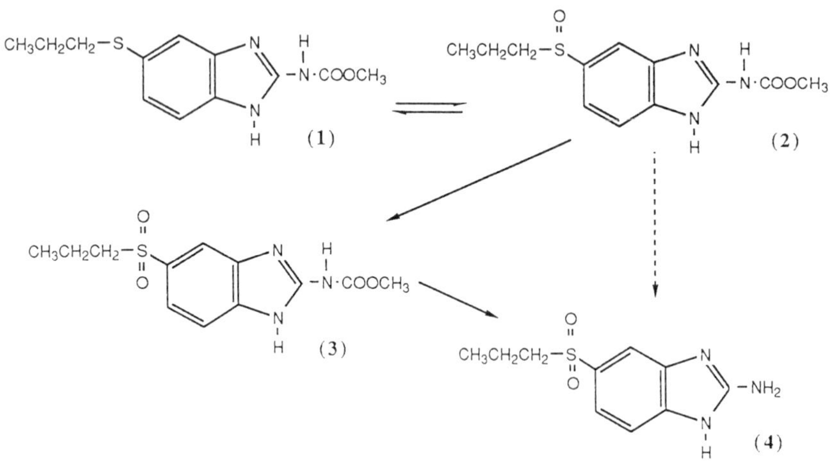

Figure 2 Major Metabolic Pathway of Albendazole (Adapted from ref.5)
where (1) = ABZ
(2) = ABZ-SO
(3) = ABZ-SO_2
(4) = ABZ-NH_2

Animal Husbandry and Dosing Procedures. Eighteen Holstein cattle (9 steers and 9 heifers) with an average weight of 226.8 Kg were selected for the present study. A rumen fistula was surgically implanted in selected cattle by a veterinarian and each fistula was exteriorized and fitted with a removable cap. Following a 3-6 week recovery period, the cattle were weighed and randomly assigned to three groups as shown in Table I. The devices were subsequently placed in the rumen of each animal via the fistula. Three cattle were selected for tissue residue analysis following 0-Day and 5-Day withdrawal. The 0-Day withdrawal point was defined as the day when total depletion of the 'Captec' device contents had occurred for the first of the three devices in each 0-D and 5-Day withdrawal groups.

Table I. 'Captec' Cattle Group Assignments

Group	*Cattle I.D.*	*Sex*	*Withdrawal Period (Days)*
	#352	F	------
	#350	F	------
	#348	F	------
I	#337	M	------
	#334	M	------
	#341	M	------
	#347	F	------
	#353	F	------
	#354	F	------
II	#336	M	0
	#343	M	------
	#335	M	------
	#345	F	0
	#349	F	5
III	#355	F	5
	#342	M	0
	#340	M	------
	#338	M	5

Group I refers to 'Captec' devices containing 1987 commercial albendazole tablets (Lot # 021389/70304-1A)
Group II refers to 'Captec' devices containing 1989 commercial albendazole tablets (Lot # 90119-28)
Group III refers to 'Captec' devices containing albendazole tablets prepared in-house (at SBAH, Lot # RCG 17813-92)

Plunger Measurement. At selected intervals, the 'Captec' devices were retrieved from the rumen through the fistula using the attached nylon string and measurements of plunger movement were taken using a pair of calipers (Figure 1). Initially, weekly measurements of plunger measurement were taken, but this frequency increased when the devices were nearing depletion in order to accurately assign the 0-Day withdrawal point. Plunger measurement required opening of the rumen to the air which affected feed consumption in some animals. Therefore, a compromise between obtaining sufficient data points and maintaining the overall animal health was required for each animal on an individual basis.

Plasma Collection. Blood samples were collected from all animals at 0, 2, 4, 6, 12, 18 and 24 hours after the devices were inserted, then once daily until Day 10. On Days 13, 18, and 21, blood samples were withdrawn again followed by weekly collections between Day 21 and Day 84. Blood samples were subsequently taken on Days 86, 88 and Day 90, and then daily until the end of the in-life study when all the devices had been depleted. Approximately 10 mL of blood was collected from all animals in VACUTAINER tubes containing heparin at each sampling point. The blood was immediately centrifuged at 2000 rpm for ten minutes and the plasma was stored frozen until extraction.

Analytical Methods

Plasma. Albendazole (ABZ), albendazole-sulfone (ABZ-SO_2), albendazole 2-aminosulfone (ABZ-NH_2, marker metabolite) and albendazole-sulfoxide (ABZ-SO), were extracted from the plasma using a perchloric acid precipitation method followed by a solid phase extraction procedure. The residue levels were quantitated using a normal or reversed phase HPLC method with fluorescence detection.

One mL of plasma was spiked with 20 µL of a 10 µg/mL solution of the internal standard, vortex-mixed for approximately one minute and let stand for one hour at room temperature. The sample was protected from sunlight during the entire process. A 0.5 mL aliquot of 6% perchloric acid was then added followed by brief vortexing, centrifugation, and removal of the supernatant. One mL of deionized water was then added to the remaining residue in the centrifuge tube. Following vortexing and centrifugation at 2000 rpm for 15 minutes, the supernatant was removed and combined with the initial sample. The pH of the combined supernatants was adjusted to 9.5 with a 10% sodium carbonate solution. The entire sample along with a 1 mL wash of the supernatant container was applied to a Waters SEP-PAK C18 cartridge which had been prewashed with 1 mL HPLC grade methanol and 1 mL deionized water. The sample was subsequently eluted with 2 ml of HPLC grade acetonitrile. The acetonitrile eluant was evaporated to dryness with a gentle stream of nitrogen and stored frozen until HPLC analysis.

The plasma extract was reconstituted in 5:1 acetonitrile/acetic acid solution just prior to HPLC analysis of ABZ, ABZ-SO_2 and ABZ-NH_2. For ABZ-SO HPLC analysis, the sample extract was reconstituted in mobile phase (see below).

Reversed phase HPLC conditions for ABZ-SO_2 and ABZ-NH_2 were as follows.

Column :	25 cm x 4.6 mm Econosphere C18 5 μ (Alltech)
Guard Column:	3 cm x 4.6 mm C-18 Brownlee Inc.
Flow rate:	2 mL/minute
Mobile phase:	60% 0.01M KH_2PO_4 : 40% acetonitrile
Injection volume:	50μL

Normal Phase HPLC conditions for ABZ-SO are as follows:

Column:	25 cm x 4.6 mm Econosphere Silica 5 μ (Alltech)
Guard Column:	3 cm x 4.6 mm Silica Brownlee Inc.
Flow rate:	2.5 mL/minute
Mobile phase:	60 : 25 : 15 : 0.1 Chloroform : Hexane : Acetic Acid: Ammonium Hydroxide
Injection volume:	70μL

Reversed phase HPLC conditions for ABZ are as follows:

Column:	25 cm x 4.6 mm Econosphere C18 5 μ (Alltech)
Guard Column:	3 cm x 4.6 mm C-18 Brownlee Inc.
Flow rate:	1.5 mL/minute
Mobile phase:	50% 0.01M KH_2PO_4 : 50% acetonitrile
Injection volume:	50μL

All HPLC analyses employed a Perkin-Elmer LS-4 fluorescence detector operating at excitation and emission wavelengths of 296 nm and 326 nm respectively.

Tissue. The liver homogenates were analyzed according to the procedure shown on Figure 3. Fraction (a) was used to determine the total marker residue level in liver. The amount of marker residue that was bound in liver tissue was obtained using fraction (b) while that of the extractable portion was determined using fraction (c), i.e.

Total marker residue in liver (a) = marker residue in bound tissue (b)
+
extractable marker residue (c)

The level of marker residue(ABZ-NH_2) was determined using the approved regulatory method entitled "Quantification of Albendazole Marker Residue in Cattle Liver Tissue Using a Fluorescence HPLC Method" (*6*).

Results

Plunger Movements and Daily Dosage Level. The 'Captec' devices exhibited a consistent uniform release rate. The average plunger movement up to 85 days was 0.7 mm/day for all groups which corresponds to a daily release of 0.2 g, or an average dose of 0.9 mg/kg. Figure 4 shows the results for heifer #348 which are typical of all animals in the study.

Plasma Residues. Plasma samples from two cattle from each group were analyzed for levels of ABZ and ABZ SO, however, neither compound was detected at any time point (detection limits of 100ppb and 25ppb respectively). The ABZ-NH2 levels were below 350 ppb throughout the study period and there was no obvious correlation between the plunger movement and the plasma residue level of ABZ-NH2. The concentrations of ABZ-SO2 were less than 90 ppb at all time points but generally averaged between 20-40 ppb for all animals. ABZ-SO2 was the only metabolite seen consistently in the plasma throughout the study duration. Its presence during the primary drug release period correlated well with the uniform plunger movement (Figure 5) and can thus be used as a monitor for proper functioning of the device when given to intact animals.

Tissue Residues. Liver samples were analyzed using the approved regulatory method for the marker, albendazole 2-aminosulfone. In this method, the liver homogenate ((a), Figure 3), was subjected to vigorous acid hydrolysis conditions, 6 N HCL for 1 hours at 110 ° C before any clean-up procedures. Using this method, the albendazole marker residue levels in the liver of 'Captec' cattle at 0- and 5-Day withdrawal were 651 ± 38 ppb and 317 ± 43 ppb, respectively. When the same method was applied to an ethyl acetate tissue extract and the remaining tissue pellet (Figure 3, (c) and (b) respectively), the data clearly indicated that nearly all of the recovered marker was released from the bound residue and was not freely extractable. The results are summarized in Table II.

Table II. Extractable vs. Bound Residues for 'Captec' Device

Withdrawal (Days)	*Extractable * Residues(%)*	*Bound * Residues(%)*	*Marker Concentration(ppb)*
0	3	97	651
5	0	100	317

* Estimates based on marker method analysis of extractable analysis

Discussion

When albendazole is administered as a single dose at 10 mg/kg b.w., albendazole sulfoxide (C_{max} 2-3 μg/mL; T_{max} 12-16 hours) and albendazole sulfone (C_{max} 1-2 μg/mL; T_{max} 18-24 hours) are routinely seen in the plasma of cattle (*1*). When ABZ was administered via the 'Captec" sustained release device, as in this study, the plasma profiles were markedly different. Only ABZ-SO2 was seen in the plasma with concentrations averaging 20-40 ppb. ABZ-SO was not observed. Thus, a reduction in the daily dose of approximately 10 fold (10mg/kg vs 0.9 mg/kg/day) led to over a 100-fold reduction in plasma metabolite levels.

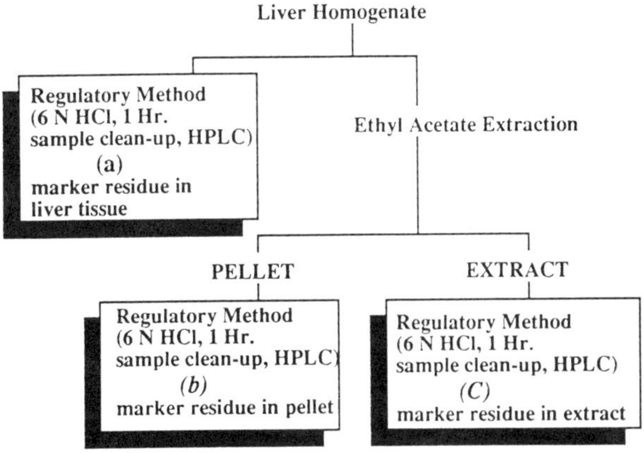

Figure 3 Flow Chart Showing the Extraction Scheme

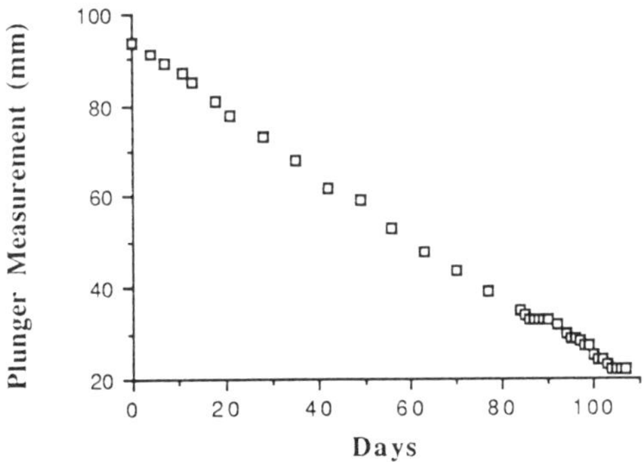

Figure 4 Plunger Movement of A 'Captec' Device

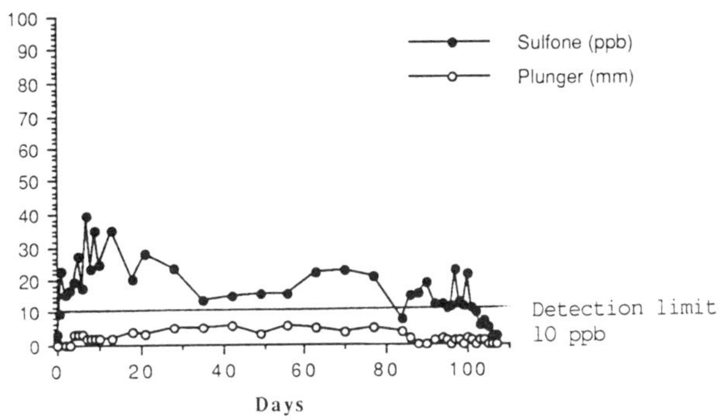

Figure 5 Plasma Sulfone Levels vs. Plunger Movement vs. Days

The ABZ tissue residue profile also shows marked contrasts between the single and sustained release dosage forms. Previous data (Wang, R.; Marker Residue Depletion Study in Liver from Cattle treated with Albendazole, SKB Applebrook Research Center Study Report, 1983) from single dose studies indicate that at 0-1 days withdrawal albendazole residues are 50% bound and 50% extractable with a marker concentration of 4500 ppb. After 4 days withdrawal, the residues are 95% bound and only 5% extractable with a marker concentration of 1200 ppb. At longer withdrawal times, the bound/extractable ratio is maintained at 95:5 while the marker depletes to less than 250 ppb at 20 days. The data generated in the present study using the 'Captec' devices indicate that nearly 100% of the albendazole residues are bound even at 0-Day withdrawal with a significantly lower marker residue of only 651 ppb.

For tissue residue monitoring, the FDA uses the concept of the Rm value which is defined as the ratio of the concentration of the designated marker metabolite to that of the total residue,

$$R_m = \frac{[\text{Marker}]}{[\text{Total Residue}]}$$

For a certain drug, the total residue is determined from a radioisotope study, and the marker concentration is determined by the approved regulatory method. Once the Rm value is known, a marker residue tolerance is established. This value is then used in field monitoring programs to ensure that no edible tissues containing drug residue above the safe concentration will reach the consumer. For albendazole, the Rm value for the 2-aminosulfone metabolite in 0.20 or 20% of the total residue. Since the safe concentration for albendazole residues in liver is 1.2 ppm (0.6 ppm in meat x 2 (food factor)), the marker tolerance has been established at 240 ppb.

The Rm concept, however, allows for only one value to be assigned for each drug and must be applied universally regardless of the method or route of drug administration (e.g. single dose or sustained release) as monitoring agencies receiving tissue sample for analysis have no information on drug treatment regimens for individual animals. The current Rm value and 28 day withdrawal time for albendazole were established using data submitted from single-dose studies.

The substantial differences (from single dose ABZ) observed in the plasma and tissue residue profiles in this study, tend to indicate that the quantitative metabolic profile may be altered in the case of the sustained-release of albendazole. If the established Rm = 0.2 factor is applied to the marker metabolite concentration of 651 ppb observed at 0-day withdrawal in this study, a total residue of 3255 ppb (651x5) would be indicated. However, considering the facts that

(a) the 2-aminosulfone is an end metabolite,
(b) plasma levels were reduced 100 fold, and
(c) tissue residues were nearly 100 % bound even at 0-Day withdrawal,

it seems unlikely that the same relationship of marker/total residue (Rm) is applicable for the sustained release of ABZ. In fact, the data suggest that the majority of the residues are present as the 2-aminosulfone. If true, then the actual Rm value would be higher and possibly be significantly greater than 20%. When subsequently used in calculations of total residue, the overall effect would be to lower the total residue (multiply marker concentration by a lower number).

The above discussion, of course, is only speculation at this point, since radioisotope data would be required to firmly establish the Rm relationship for the sustained release dosage form. Generating this type of data is not straightforward, however, due to the large amount of radiolabelled drug required and the problems inherent in maintaining cattle in metabolism cages for up to 90 days or longer.

A potential alteration of the Rm value for a sustained release dosage form of ABZ has regulatory consequences as well. Since the initial Rm = 0.2 was established for albendazole as a single dose, this value must be applied to all dosage forms and routes of administration. The regulatory method cannot distinguish between cattle given a single dose of ABZ or those dosed via 'Captec'. The data from the present study indicate that it is quite possible that cattle dosed with albendazole via the 'Captec' device would qualify for 0-Day withdrawal. Unfortunately, even if the radioisotope data were available at this time to definitively determine the Rm for sustained release ABZ, it could not be used to establish a new tolerance for the drug as long as both dosage forms remained commercial products.

Acknowledgments. The authors would like to thank C. Gombatz for her advices and dicussions, and D. Hallman, F. Hallman, R. Kozan, J. Rizzo, S. Shuman, P. Stone, R. Supplee, E. Warrakeh and Dr. L. Weerasinghe for their technical support.

Literature Cited

1. Glanzer, K.; Pfeiffer, H.; Wetzel, H. Acta Pharm. Technol. **1988,** *34*, pp108-114
2. Prichard, R.K.; Hennessy, D.R.; Steel, J.W; Vet. Parasitol. **1978,** *4*, pp 309-315
3. Kwan, L.C.; Gyurik, R.J.; Freeman, J. F.; Chimes, N.; Ritch, G. M.; Theodorides, V.J. J. of Control Release **1988,** *8*, pp 31-38
4. Gyurik, R.J.; In *Rumen Retention Devices*; Tyle, P., Ed.; Drug Delivery Devices; Marcel Dekker: New York and Basel, Vol. 20; pp 549-561
5. Gyurik, R.J.; Chow, A.W.; Zaber, B; Brunner, E.L.; Miller, J. M.; Villani, A.J. Villani; Petka, L. A.; Parish, R.C. Drug Metab. Dispos. **1981,** *9*, pp 503-508
6. Wishousky, T.; Wang, R. W. *Quantification of Albendazole Marker Residue in Cattle Liver Tissue Using a Fluorescence HPLC Method* On display in 3rd Additives Manuel, FOI Public Information Room, Room 12A-30, Food and Drug Adminstration, 5600 Fishers Lane, Rockville, MD 20857

RECEIVED March 13, 1992

Chapter 11

Tilmicosin in Cattle

Metabolism and Tissue Residues

A. L. Donoho, T. D. Thomson, and D. D. Giera

Lilly Research Laboratories, A Division of Eli Lilly and Company, Greenfield, IN 46140–0708

Tilmicosin, (20-deoxo-20-(3,5-dimethylpiperidin-1-yl)-desmycosin) is a new macrolide antibiotic which is being developed for treatment of bovine respiratory disease by single subcutaneous injection into cattle. Excretion, tissue residue, and metabolism studies were conducted with ^{14}C-tilmicosin-treated cattle and rats. Radioactivity was excreted primarily in the feces after parenteral dosing of cattle and oral dosing of rats. Among edible tissues from treated cattle, liver and kidney contained the highest concentrations of radioactivity. The most abundant metabolite was N-desmethyl tilmicosin. Comparative metabolism studies with cattle and rats indicated that the pattern of metabolism was the same for both species.

Tilmicosin, Figure 1, is a new semisynthetic antibiotic derived from the macrolide antibiotic tylosin. The synthesis and antibacterial activity of tilmicosin have been described by Debono, et al. and Kirst, et al. (*1-4*). Tilmicosin has *in vitro* activity against a variety of Gram positive and Gram negative bacteria, as well as mycoplasma species. It is effective for treatment of bovine respiratory disease caused by *Pasteurella haemolytica* when administered as a single subcutaneous injection (*5, 6*). This paper describes the excretion, tissue residue pattern, and metabolism of tilmicosin when injected into cattle, and also gives comparative metabolism data from rats which were dosed orally.

Materials and Methods

Tilmicosin. The macrolide antibiotic tylosin is a fermentation product of *Streptomyces fradiae*. Tylosin is converted to desmycosin by acid hydrolysis to remove the mycarose ring and tilmicosin is produced from desmycosin by reductive amination of the C-20 aldehyde with 3,5-dimethylpiperidine (*1*). Tilmicosin is a mixture of diastereomers having the methyl groups of the piperidine ring approximately 85% cis and 15% trans.

Labeled Compound. Two different ^{14}C-tilmicosin preparations were used in these studies. One was piperidine labeled ^{14}C-tilmicosin which was prepared by reacting (2,6^{14}C) 3,5-dimethylpiperidine (California Bionuclear, Sun Valley, CA) with unlabeled desmycosin. This material is designated as Pip-^{14}C-tilmicosin. The second was an equimolar mixture of ^{14}C-tilmicosin labeled in the macrolide ring (87% of the

0097–6156/92/0503–0158$06.00/0

^{14}C) and ^{14}C-tilmicosin labeled at positions 3 and 5 in the dimethylpiperidine ring (13% of the ^{14}C) (*7*). This material is designated as Eq-^{14}C-tilmicosin. Macrolide labeled ^{14}C-tilmicosin was prepared by reacting ^{14}C-desmycosin with unlabeled 3,5-dimethylpiperidine. ^{14}C-Desmycosin was prepared from ^{14}C-tylosin which was produced by fermentation using 1-^{14}C-propionate as substrate. Positions of labeling of the macrolide ring are inferred from stable isotope synthesis studies (*8*) and are shown in Figure 1.

Dosing Procedures. Cattle were treated by a single subcutaneous injection of tilmicosin in aqueous propylene glycol solution. The tilmicosin concentration of the dosing solution was 300 mg/mL. For the radiolabeled studies, cattle were kept in individual metabolism stalls. Feces and urine were collected daily where appropriate. Tissues were collected at slaughter, and then ground. Male and female Fisher strain 344 rats were dosed orally with Eq-^{14}C-tilmicosin aqueous solution for three consecutive days at a dose of 20 mg tilmicosin per kg body weight. Excreta were collected daily and pooled by sex. Three days after completion of dosing the rats were killed and livers were collected for assay. All samples from cattle and rats were held frozen until assayed.

Assay Procedures. Tissues, feces, urine samples, and sample fractions were assayed for radioactivity (RA) by liquid scintillation counting essentially as described by Magnussen et al. (*9*). Tissues were assayed for parent tilmicosin by methanol extraction, purification by liquid-liquid partitioning, and measurement by high performance liquid chromatography (HPLC) on a reversed phase phenyl column with detection by UV absorption at 280 nm. Elution was with a nonlinear gradient of mobile phase consisting of water, acetonitrile, and dibutyl ammonium phosphate at pH 2.5.

For characterization of RA in tissues and excreta the following general extraction and purification scheme was developed. Samples were extracted with 80:20 methanol:water and the extract was diluted with a volume (usually 0.6 v/v) of 10% aqueous sodium chloride solution. The diluted extract was partitioned at ca. pH 6 with CCl_4 and then adjusted to ca. pH 8. The tilmicosin and metabolites were extracted into 1:1 $CCl_4/CHCl_3$ for further purification or evaluation. Thin-layer chromatography (TLC) was conducted with silica gel plates developed in 100:15:2 $CHCl_3$:methanol:ammonium hydroxide. Autoradiograms were prepared by overlay of the plates with X-ray film.

Balance-Excretion Studies

Balance-excretion studies were conducted with four steers in three experiments. In the first study, a steer was given a single injection of Pip-^{14}C-tilmicosin at a dosing level of 30 mg/kg body weight to obtain preliminary data on excretion. After a 15 day collection period, 19% of the dose was recovered in urine, 72% in the feces, and approximately 5% in the carcass. These data indicated feces to be the primary excretory pathway. In the second study, a steer was dosed with 20 mg/kg Pip-^{14}C-tilmicosin and collections were made for 21 days. In the third study, two steers were dosed with 10 mg/kg Eq-^{14}C-tilmicosin and collections were made for 14 days. Rates of excretion and total recovery of RA from these two studies are compared in Figure 2. The steer dosed with Pip-^{14}C-tilmicosin excreted 68% of the dose in the feces and 24% in the urine. The two steers dosed with Eq-^{14}C-tilmicosin excreted 56 and 52% in feces and 18 and 21% in urine, respectively. The majority of the radioactive dose was recovered in both experiments. However, recovery was higher from the steer dosed with Pip-^{14}C-tilmicosin (92%) than the two steers dosed with Eq-^{14}C-tilmicosin (74%).

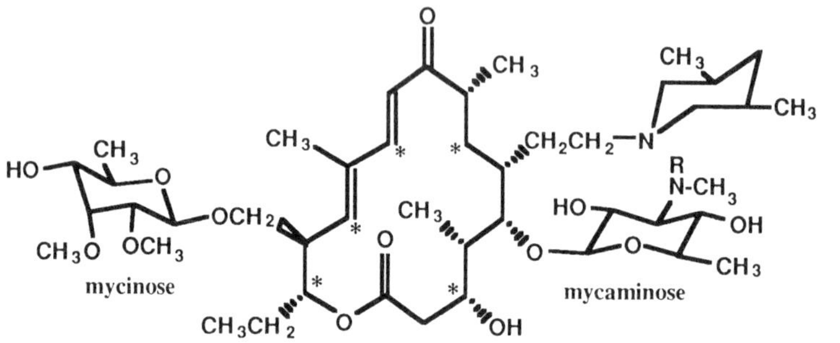

Figure 1. Structure of tilmicosin ($R=CH_3$) and metabolite T-1 (R=H). Note: Positions of labeling of macrolide ring are shown by (*). Pip-^{14}C-tilmicosin was labeled in the 2,6 positions of the piperidine ring and Eq-^{14}C-tilmicosin was labeled in the macrolide ring and the 3,5 positions of the piperidine ring.

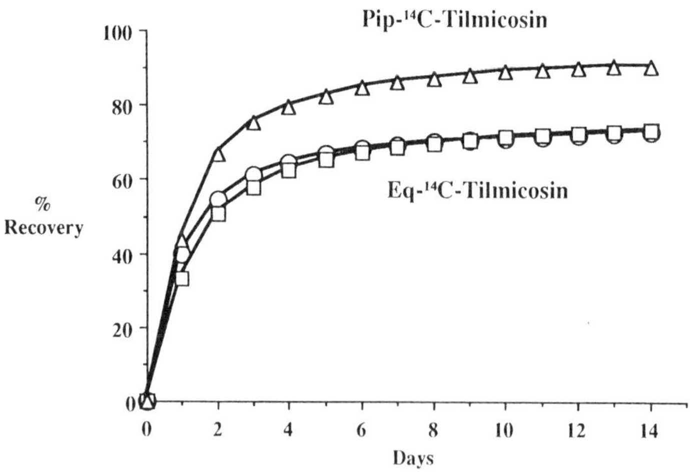

Figure 2. Excretion of radioactivity by a steer dosed with 20 mg/kg Pip-^{14}C-tilmicosin (upper curve) and by two steers dosed with 10 mg/kg Eq-^{14}C-tilmicosin (lower two curves).

Tissue Residue Studies

Three tissue residue studies were conducted to characterize the pattern of residues in cattle treated with tilmicosin. In the first study, four steers and one bull calf weighing 157 to 202 kg were injected subcutaneously with 20 mg/kg Pip-^{14}C-tilmicosin. One calf was euthanized 3 days after dosing, two were euthanized at 21 days, and two at 56 days after dosing. Total RA was measured in muscle, liver, kidney, fat, injection site, lung, and bile. In addition, blood was collected from one calf at 4, 8, 24, 48, 72, and 96 hours after dosing. Blood plasma RA assays gave values of 2.0, 1.5, 0.5, 0.2, 0.17, and 0.11 ppm for the above sampling times. Microbiological assay of the plasma samples for tilmicosin indicated that most of the plasma RA was parent tilmicosin. Results from assay of tissues and bile are shown in Table I. At the earliest withdrawal, liver, kidney, and injection site contained the highest concentrations of RA. Muscle and fat contained relatively little residue while lung and bile had intermediate values. The rate of RA decline indicated liver to be the tissue with the highest and most slowly depleting residue.

Table I. Residues of Radioactivity (Mean PPM) in Tissues of Cattle Injected with 20 mg/kg of Pip-^{14}C-Tilmicosin

	Withdrawal (Days)		
Tissue	3	21	56
Liver	36.00	5.51	1.05
Kidney	39.20	2.31	0.23
Muscle	2.00	NDR	NDR
Fat	1.27	<0.10	NDR
Lung	14.36	0.93	<0.10
Inj Site	81.63	5.25	0.21
Bile	24.74	0.40	NDR

NDR = No Detectable Residue

The second tissue residue study was conducted after the dose was established at 10 mg/kg body weight. Twelve cattle, seven steers and five heifers, weighing 182 to 217 kg, were dosed at 10 mg/kg with Eq-^{14}C-tilmicosin. One steer and one heifer were euthanized 3 days after treatment. Similar pairs were euthanized at 14 days and 56 days. Groups of two steers and one heifer were euthanized at each withdrawal time of 28 and 42 days. Muscle, liver, kidney, fat, injection site, and bile samples were assayed for total RA and selected tissues were assayed for parent tilmicosin. Results are shown in Table II. The magnitude of residue at 3 days was comparable to the first study, considering the two-fold difference in dose, and the pattern of decline was similar. Liver contained slightly higher residues of RA at longer withdrawal times when dosed with Eq-^{14}C-tilmicosin. This was probably due to metabolism of the ^{14}C in the macrolide ring and incorporation into more slowly depleting compartments of RA. By 28 days, parent tilmicosin was only a small percentage of total liver RA and other edible tissues had small residues compared to liver. Assay for tilmicosin in liver was a reliable means of monitoring the residue status of treated cattle. Therefore, parent tilmicosin is considered to be the marker compound for analysis. Nonextractable liver RA increased as a percentage of total at longer withdrawals. A comparison of total liver residue, parent tilmicosin and nonextractable residue is shown in Figure 3. HPLC assays of muscle, including injection site, gave values of approximately half of the RA as parent tilmicosin.

Table II. Radioactivity (RA) and Tilmicosin (Til) in Tissues of Cattle Injected with 10 mg/kg of Eq-^{14}C-Tilmicosin

Withdrawal (Days)		Mean ppm					
		Liver	Kidney	Muscle	Fat	Inj Site	Bile
3	RA	19.44	18.09	0.41	0.24	73.53	9.54
	Til	7.11		0.18		42.0	
14	RA	11.63	2.51	0.09	0.05	13.82	<1.00
	Til	1.99		<0.05		8.3	
28	RA	5.74	0.59	<0.05	0.03	5.07	<0.20
	Til	0.38				2.6	
42	RA	3.52	0.27	NDR	<0.04	0.94	<0.10
	Til	<0.10					
56	RA	2.72				0.33	NDR
	Til	<0.05					

NDR = No Detectable Residue

In the third tissue residue study, twelve cattle, eight steers, and four heifers, weighing approximately 200 kg, were dosed with 10 mg/kg unlabeled tilmicosin. Groups of 2 steers and one heifer were euthanized at 14, 28, 35, and 42 days after treatment. Muscle, liver, kidney, fat, and injection site samples were collected and assayed for parent tilmicosin. The mean residue concentrations are shown in Table III.

Table III. Tilmicosin Tissue Residue Decline Study in Cattle Injected with 10 mg/kg Unlabeled Tilmicosin

Withdrawal (Days)	Tilmicosin Concentration (ppm)				
	Liver	Kidney	Muscle	Fat	Inj Site
14	0.92	0.94	<0.05	<0.05	18.94
28	0.26	0.14	<0.05	<0.05	2.92
35	0.18	0.11	<0.05	NS	0.78
42	<0.10	<0.10	NS	NS	0.29

Assay Limit = 0.05 ppm
NS = No Sample

Metabolism Studies

Metabolism studies were conducted on cattle urine, feces, and tissue samples to determine the pattern of metabolism and characteristics of the radioactive residues. In addition, comparative metabolism work was done on rat urine, feces, and liver. All of the metabolism work described is from animals treated with Eq-^{14}C-tilmicosin, since this dosing material was the most uniformly labeled.

Metabolites in Tissues. Liver was the cattle tissue with the highest and most slowly depleting residue of RA. Therefore, the greatest effort was directed toward characterization of liver residues. A pool of 30 g of liver from the two 3-day withdrawal cattle and a second pool of 30 g from the three 28-day withdrawal cattle from the second tissue residue experiment were processed by the general scheme described above. The distribution of RA for the 3- and 28-day samples was: nonextractable RA 17% and 36%, respectively, CCl_4 <1%, aqueous methanol 10% or less, and $CCl_4/CHCl_3$ fractions 75 and 54%, respectively. HPLC fractionation was done on the $CCl_4/CHCl_3$ fractions and the radiochromatograms are shown in Figure 4. There were only three primary peaks of RA in the extractable fraction. These were, parent tilmicosin, metabolite T-1, which eluted just before tilmicosin, and compound T-2, which eluted after tilmicosin. Metabolite T-1 constituted 16% and 8% of the liver RA at 3- and 28-days, respectively. Compound T-2 constituted 9% and 21% in the same samples. Metabolite T-1 was later identified as N-desmethyl tilmicosin. Compound T-2 was found to be a minor impurity in the dosing material, rather than a true metabolite. Characterization of T-1 and T-2 are described in a later section. Residues in livers of the 42- and 56-day withdrawal cattle were too low for this type of fractionation.

Fractionation of RA in a kidney sample from the 3-day withdrawal animals was conducted similarly to that described for liver. The distribution was: nonextractable 21%, aqueous methanol 14% and $CCl_4/CHCl_3$ fraction 65%. The radiochromatogram from HPLC analysis is shown in Figure 5. Most of this fraction was parent tilmicosin, with a small amount of T-1 and polar radioactivity. In contrast to liver, kidney had very little T-2.

Comparison of total RA and HPLC muscle and injection site data indicated that approximately half was parent tilmicosin (44% for the 3-day muscle, and 50 to 60% for injection sites). No further fractionation of muscle RA was done. The unlabeled tissue residue study demonstrated that tilmicosin in muscle was <0.05 ppm at 14 days withdrawal.

Fat residues of RA were <0.05 ppm by 14 days withdrawal, demonstrating that there was no accumulation of residue in the fat tissue. No further characterization was conducted on fat tissues.

Metabolites in Excreta. Urine and feces from treated cattle were processed by the general fractionation scheme. Samples of urine at pH 6 were extracted with CCl_4 and then again at pH 8 with CCl_4 and $CCl_4/CHCl_3$. The latter two urine extracts combined contained 90% of the RA and was found to be primarily parent tilmicosin when assayed by HPLC, giving an estimate of 74% tilmicosin in Day 1 urine. Metabolite T-1 was also present in small quantities in urine. The major extractable fraction from feces, which included a $CHCl_3$ extraction following the $CCl_4/CHCl_3$ extraction contained approximately 79% of the RA. This fraction was evaluated by TLC and estimates of tilmicosin and metabolites were made by counting segments scraped from the plate. Tilmicosin and metabolite T-1 each accounted for about 22% of the fecal RA. Two minor metabolite zones together constituted 18%. One of these was later identified as T-3 (<10% of total RA) and the other zone was multiple compounds.

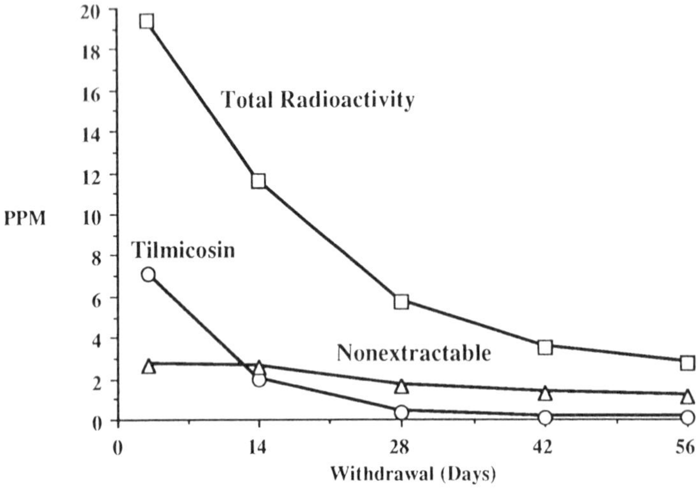

Figure 3. Radioactivity in different fractions from livers of cattle dosed with 10 mg/kg Eq-^{14}C-tilmicosin.

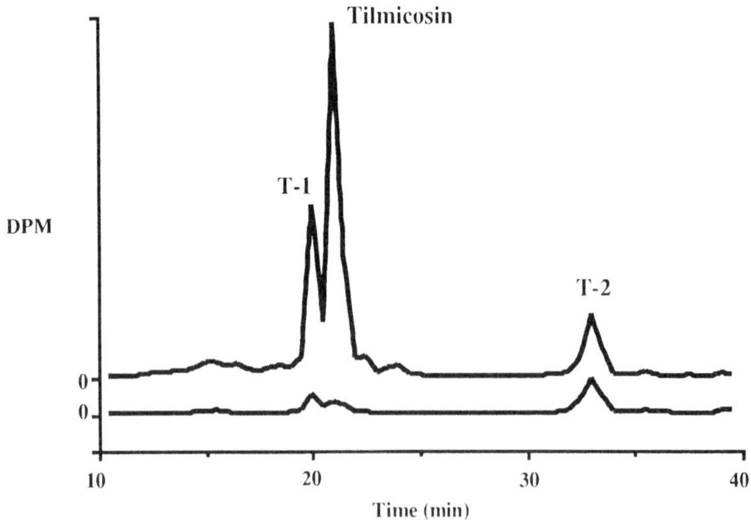

Figure 4. Radiochromatograms of liver extracts from cattle dosed with 10 mg/kg Eq-^{14}C-tilmicosin, 3-day withdrawal (top), and 28-day withdrawal (bottom).

Comparative Metabolism. One of the important questions in evaluation of human food safety of animal drugs is whether the drug metabolites present in tissues of food-producing animals have been evaluated for safety. This assessment is made through comparative metabolism studies to show the presence in excreta or tissues of toxicological test species of the same metabolites which appear in edible animal tissues. The primary comparison for use of tilmicosin in cattle was the residue composition of cattle liver compared to excreta or tissues of rats which were treated with ^{14}C-tilmicosin. The three characterized components of the liver residue were tilmicosin, metabolite T-1, and compound T-2. Both tilmicosin and T-2 were present in the dosing material which was used in the toxicology studies. Therefore, the primary question was whether metabolite T-1 was produced by rats. An HPLC radiochromatogram from an extract of rat liver is shown in Figure 6. Metabolite T-1 and tilmicosin were the primary radioactive liver components. Rat feces contained most of the excreted radioactivity from the treated rats. Therefore, rat feces extracts were prepared and HPLC and TLC comparison of cattle liver and rat feces extracts also confirmed the presence of T-1 as a rat metabolite.

Identification of Metabolites T-1 and T-3. Metabolite T-1 was isolated from a total of 1.6 kg of cattle feces by the general extraction and partitioning steps described above. In the first preparation, the $CCl_4/CHCl_3$ fraction was processed by preparative silica gel column chromatography and silica gel preparative TLC to yield three TLC zones. One was metabolite T-1 and the second zone was called metabolite T-3. The third zone appeared to consist of multiple compounds in only small quantities and was not processed further. The T-1 isolate from the TLC plate was characterized by proton NMR spectrometry which indicated that the difference in structure between T-1 and tilmicosin was on the mycaminose ring. The FAB mass spectrum was inconclusive. Subsequently, a second T-1 isolate was purified by additional partitioning and by preparative HPLC on a C-18 reversed phase column. The FAB mass spectrum gave a MH+ ion of 855 m/z. High resolution MS of the major peak gave a composition of $C_{45}H_{79}N_2O_{13}$. There was no m/z 174 ion for the mycaminose fragment. However, there was a m/z 160 ion equivalent to mycaminose minus CH_2. Therefore, T-1 is presumed to be tilmicosin minus one of the mycaminose N-methyl groups (Figure 1).

Metabolite T-3 was a minor metabolite isolated during the T-1 purification. Field desorption MS indicated a molecular weight of 841. FAB MS indicated the dimethylpiperidine to be intact, but the mycaminose ring to be altered. The uneven molecular weight indicated the loss of one nitrogen. Therefore, the proposed structure for T-3 is the replacement of $N(CH_3)_2$ on the mycaminose ring with OH.

Identification of Compound T-2. The metabolism work on liver suggested that Compound T-2 in liver was from the dosing material rather than from metabolism. A quantity of T-2 was isolated from technical tilmicosin by countercurrent partitioning and preparative HPLC on a C-18 reversed phase column. Extensive NMR and MS analyses were conducted. The structure of T-2 that was deduced from this work is that it is a dimeric derivative of tilmicosin. A description of this work is beyond the scope of this paper and will be published later.

The presence of increased concentrations of compound T-2 relative to tilmicosin in livers from cattle which were injected with tilmicosin has been noted above. This presence of T-2 beyond trace levels was not observed in livers of orally treated rats (Figure 6).

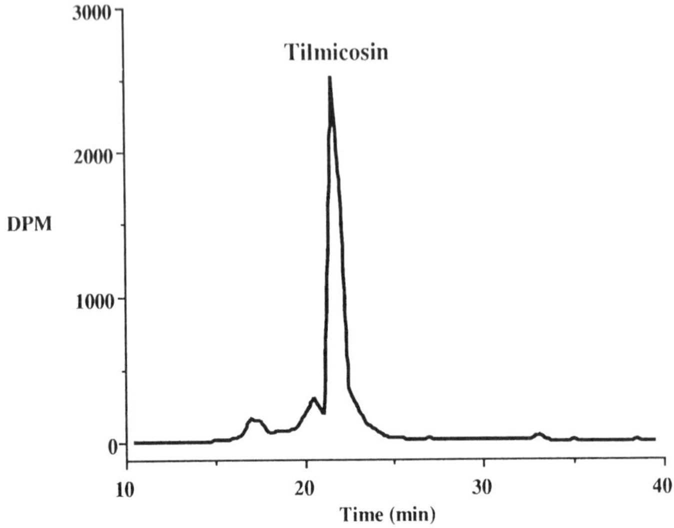

Figure 5. Radiochromatogram of kidney extract from cattle dosed with 10 mg/kg Eq-^{14}C-tilmicosin and euthanized at 3-day withdrawal.

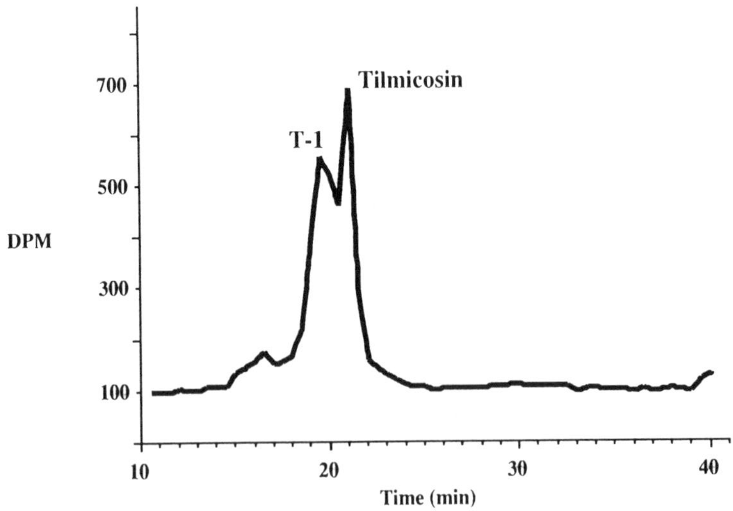

Figure 6. Radiochromatogram of rat liver extract.

Conclusions

Cattle dosed parenterally with ^{14}C-tilmicosin excreted most of the dose in feces. Recovery of the dose was higher in animals dosed with ^{14}C-tilmicosin labeled in only the piperidine ring than those dosed with ^{14}C-tilmicosin labeled in both the macrolide and piperidine rings. This may be due to metabolism of the macrolide ring. In cattle dosed with 10 mg/kg, mean parent tilmicosin residues declined to less than 1 ppm in the four edible tissues by 28 days after treatment. Liver was the target tissue, ie. the tissue with the highest and most slowly depleting residues. Assay for parent tilmicosin in liver was a suitable way to monitor the residue status of treated cattle. Liver contained only one metabolite with a concentration high enough for characterization. This was proposed to be N-desmethyl tilmicosin on the basis of NMR and mass spectral data. A small quantity of compound T-2, a dimeric derivative of tilmicosin which is a minor component of tilmicosin, was present in liver of treated cattle. It was not found to accumulate in livers of rats which were dosed orally.

Acknowledgments

The authors express appreciation to G. E. Babbitt for NMR analyses, J. L. Occolowitz for MS analyses, and to D. E. Ruggles and G. W. Bewley for technical assistance.

Literature Cited

1. Debono, M.; Willard, K. E.; Kirst, H. A.; Wind, J. A.; Crouse, G. D.; Tao, E. V.; Vicenzi, J. T.; Counter, F. T.; Ott, J. L.; Ose, E. E.; Omura, S. *J. Antibiotics* **1989**, *42*, 1253-1267.
2. Kirst, H. A.; Willard, K. E.; Debono, M.; Toth, J. E.; Truedell, B. A.; Leeds, J. P.; Ott, J. L.; Felty-Duckworth, A. M.; Counter, F. T.; Ose, E. E.; Crouse, G. D.; Tustin, J. M.; Omura, S. *J. Antibiotics* **1989**, *42*, 1673-1683.
3. Kirst, H. A.; Ose, E. E.; Toth, J. E.; Willard, K. E.; Debono, M.; Felty-Duckworth, A. M.; Pekarek, R. S. *J. Antibiotics* **1988**, *41*, 938-948.
4. Kirst, H. A.; Toth, J. E.; Debono, M.; Willard, K. E.; Truedell, B. A.; Ott, J. L.; Counter F. T.; Felty-Duckworth, A. M.; Pekarek, R. S. *J. Med. Chem.* **1988**, *31*, 1631-1641.
5. Ose, E. E.; Tonkinson, L. V. *Veterinary Record* **1988**, *123*, 367-369.
6. Gorham, P. E.; Carroll, L. H.; McAskill, J. W.; Watkins, L. E.; Ose, E. E.; Tonkinson, L. V.; Merrill, J. K. *Canadian Vet. J.* **1990**, *31*, 826-829.
7. Crouse, G. D. and Terando, N. H. *J. Labelled Compounds and Radiopharmaceuticals* **1989**, *27*, 465-471.
8. Omura, S.; Takeshima, H.; Nakagawa, A.; Miyagawa, J.; Piriou, F.; Lukas, G. *Biochemistry* **1977**, *16*, 2860-2866.
9. Magnussen, J. D.; Dalidowicz, J. E.; Thomson, T. D.; Donoho, A. L. *J. Ag. Food. Chem.* **1991**, *39*, 306-310.

RECEIVED March 20, 1992

Chapter 12

Ametryn in Rats, Lactating Goats, and Laying Hens

Metabolic Fate

Jinn Wu[1], Dave D. W. Liu[1], Robert A. Robinson[1], William Itterly[2], and Thomas M. Capps[2]

[1]XenoBiotic Laboratories, Inc., P.O. Box 3205, Princeton, NJ 08543–3205
[2]Agricultural Division, Ciba-Geigy Corporation, P.O. Box 18300, Greensboro, NC 27419

[^{14}C]-Ametryn was administered to rats, lactating goats, and laying hens. Excreta from all three species, daily milk from the goats, and egg samples from the hens were collected. In addition, selected tissues were taken following animal sacrifice. All samples were analyzed to determine the total radioactive residues, and metabolites were identified and quantified in excreta and selected tissues. The results showed that, in all three species, the majority of the administered dose was eliminated in excreta. Residues were observed in milk, eggs, tissues, and organs; liver and kidney showed the highest total radioactive residue levels. Metabolites in the rat were isolated, purified, and identified by thin-layer chromatography, high performance liquid chromatography, and various mass spectroscopic techniques. Ametryn appears to undergo extensive metabolic transformation by *N*-dealkylation, oxidation/hydroxylation, and conjugation (with sulfate, glutathione derivatives, and glucuronic acid). The results obtained from the rat were compared with those obtained from the ruminants and poultry.

Animal metabolism provides both a scientific and a regulatory contribution to the data base needed to register or re-register an agrochemical. Absorption, distribution, metabolism, and excretion (ADME) studies in the rat are required in the overall toxicological assessment of a chemical's safety. Metabolism studies in livestock are needed when pesticide use, either by direct treatment of animal feed items or by premises application, can result in residues in milk, meat, or eggs.

The primary focus of a rat metabolism study is the identification of metabolites in excretory products (urine and/or feces); and in a ruminant (goat or cow) or laying hen study, the primary focus is the qualitative nature of the residues (NOR) in milk, meat, and tissues. Analysis of excreta from livestock, though not critical from a regulatory perspective, can often aid in tissue identification by contributing an alternate source of an important metabolite in greater abundance than can be obtained from the tissue itself. Isolation and identification of a "marker" metabolite from goat, hen, or, in many instances, rat excreta, followed by a thorough chromatographic comparison with the tissue metabolite, can be handled in a more timely manner with less analytical frustration.

0097–6156/92/0503–0168$06.50/0

[14C]-Ametryn [2-(ethylamino)-4-(isopropylamino)-6-(methylthio)-*S*-triazine], uniformly labeled on the triazine ring, is a selective herbicide registered for control of broad-leaved and grassy weeds. (*See chemical structure below.*) To fulfill the re-registration requirement of the Pesticide Assessment Guidelines, the metabolic fate of the titled compound in three different animal species, namely, rats, lactating goats, and laying hens, was investigated.

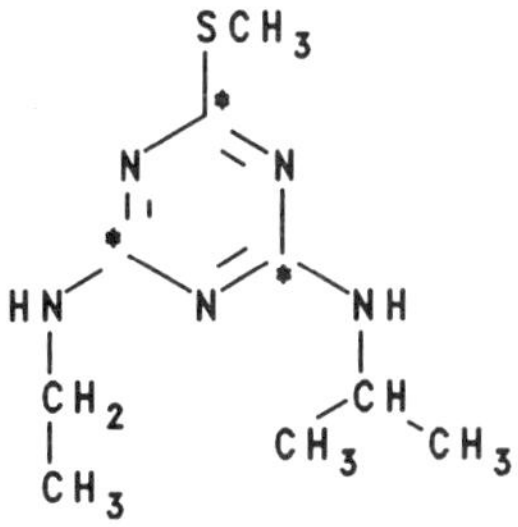

*Denotes position of ^{14}C label

[14C]-Ametryn

Study Design

Rat metabolism work was initiated first on the premise that one or more metabolites might be present in excreta, which might be useful in subsequent goat and hen studies. A preliminary rat metabolism study was conducted (1) to generate excreta samples in order to isolate and identify unknown metabolites and (2) to assess the possibility of carbon dioxide elimination. A more detailed and definitive study was subsequently initiated according to U.S. Environmental Protection Agency (EPA) guidelines. Detailed information on the dosing regimens for rats, goats, and hens is summarized in Table I.

Table I. Ametryn Dosing Summary

Animal	*Dosing Level (mg/kg)*	*Type*
Rat[a]	0.5	Single Oral Low Dose Multiple Oral Low Dose Single Intravenous Dose
	200	Single Oral High Dose
Goat[b]	1.5 (50 ppm)[c]	Multiple Oral Dose
Hen[b]	3 (50 ppm)[c]	Multiple Oral Dose

[a]Dosing followed by 7-day depuration.
[b]Animal sacrificed within 24 hours after the final dosing.
[c]Estimated dietary exposure.

A total of eight rats in two different dosing groups (two males and two females in each group), i.e., single oral low dose (0.5 mg/kg of rat body weight) and single oral high dose (200 mg/kg), were used in the preliminary study. In the Federal Insecticide, Fungicide, and Rodenticide Act (FIFRA) rat study, a total of 44 rats were utilized. Ten rats (five per sex) were used in each of the four treated groups; i.e., single oral low dose (0.5 mg/kg), multiple oral low dose (0.5 mg/kg), single oral high dose (200 mg/kg), and intravenous dose groups (0.5 mg/kg). In addition, four rats (two per sex) were used as controls. Sample collection periods were 4 days for the preliminary study and 7 days for the definitive study.

For the goat metabolism study, three lactating goats (one control and two treated) were dosed for a period of 3 days. The goats were orally administered approximately 75 mg/day (~50 ppm dietary intake) of [^{14}C]-ametryn for 3 consecutive days.

For the hen metabolism study, 15 laying hens (10 treated and 5 controls) were orally administered approximately 6 mg/day (~50 ppm dietary intake) of [^{14}C]-ametryn for 3 consecutive days.

Excreta from all three species, as well as daily milk and egg samples, were collected. Also, selected tissues were taken following animal sacrifice.

Total Material Balance and Total Radioactive Residues

Table II summarizes the total recovery of the administered dose, which is expressed as a percentage of the total administered dose, in all three species of test animals. Rats excreted nearly comparable levels in the urine and feces, with about 2% of the dose remaining in the carcass and tissues within 7 days of the radiochemical dose. A summary of the total recovery of the administered dose in goats and hens is also given in Table II. The ^{14}C recovered in goat urine, feces, daily milk, and tissues was approximately 61%, 8%, 1%, and 6%; the radiocarbon recovery profile in hen excreta, egg, and organs/tissues was similar, namely, 73%, <0.1%, and 5%, respectively.

Table II. Mass Balance and Distribution of Ametryn and Its Metabolites in the Rat, Lactating Goat, and Laying Hen

	Percent of Dose		
Matrix	*Rat*	*Goat*	*Hen*
Urine	54	61	73[a]
Feces	39	8	-
Milk	-	1	-
Eggs	-	-	<0.1
Tissues	2	6	5

[a]Excreta (combined urine and feces).

As shown in Figure 1, urinary excretion reached a plateau within the first 24 hours. Fecal elimination was delayed, indicating a possible recycling of metabolites through the rat's enterohepatic circulatory system. Some differences in behavior may exist between male and female animals; however, very little differences were observed among the various dose groups. Greater than 95% of the administered dose was eliminated 7 days after dosing.

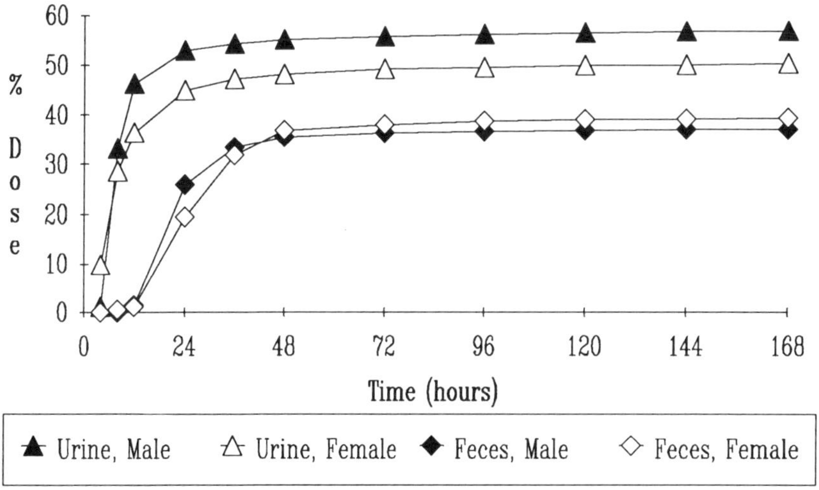

Figure 1. Ametryn Rat Metabolism Excretion Pattern

Table III summarizes the total radioactive residues (TRR) in comparable tissues of all three species dosed at approximately the same rate. Radiocarbon levels in rat tissues were low and reflected the potential effect of a 7-day depuration period. Goat and hen tissue residues were consistently higher due to the effect of a multiple dose regimen and a depuration interval of less than 24 hours.

Table III. Tissue Residues

Matrix	*Rat*[a]	*Parts per Million (μg/g)* *Goat*[b]	*Hen*[c]
Liver	0.30	2.8	4.9
Kidney	0.03	3.0	3.5
Muscle	0.01	0.1	0.5
Fat	<0.01	0.1	0.2
Milk	-	1.0	-
Egg	-	-	0.2

[a]Dose level 0.5 mg/kg.
[b]Dose level 1.5 mg/kg.
[c]Dose level 3.0 mg/kg.

Isolation and Identification of Metabolites From the Rat

As previously mentioned, rat urine and feces provided adequate sources for metabolite isolation and identification. Chromatography showed a similar product profile; however, metabolite purification from urine was selected for the convenience and ease of sample processing. Figure 2 shows comparative HPLC chromatograms of ametryn reference standards and rat urine. Under these conditions, the parent compound (reference standard No. 1, code CG-1) showed a retention time (R_t) of 31.58 min. However, a majority of the urinary metabolites did not match the available reference standards, necessitating detailed isolation and structure elucidation efforts. Figure 3 illustrates the isolation flowchart for ametryn rat metabolites. Rat urine was passed through a 1 gram C_{18} cartridge and eluted with 3 ml of water (Fraction A), followed by 2 ml of 10% acetonitrile (ACN) in water (Fraction B) and 7 ml of methanol (MeOH; Fraction C). A combination of various chromatographic techniques, such as preparative thin-layer chromatography (TLC, either normal phase or reverse phase, or both) and preparative high performance liquid chromatography (HPLC, using various columns), were then applied in order to achieve the highest purity of unknown metabolites for spectroscopic identification. A large number of isolated metabolites were chromatographically compared with known metabolite reference standards. Unknown metabolites were submitted for mass spectroscopic analyses. Depending on the nature and polarity of the metabolites isolated, one or more of the following mass ionization techniques were applied to provide the information necessary for chemical structure elucidation: direct insertion probe (DIP)/mass spectrometry (MS), gas chromatography (GC)/MS (electron impact [EI] or chemical ionization [CI]), thermal spray (TSP)/liquid chromatography (LC)/MS, fast atom bombardment (FAB)/MS (positive or negative mode), and desorption chemical ionization (DCI)/MS (positive or negative mode). A total of 13 primary metabolites were identified from the high dose rat urine (Table IV) employing the procedures discussed above.

Ametryn and Reference Standards

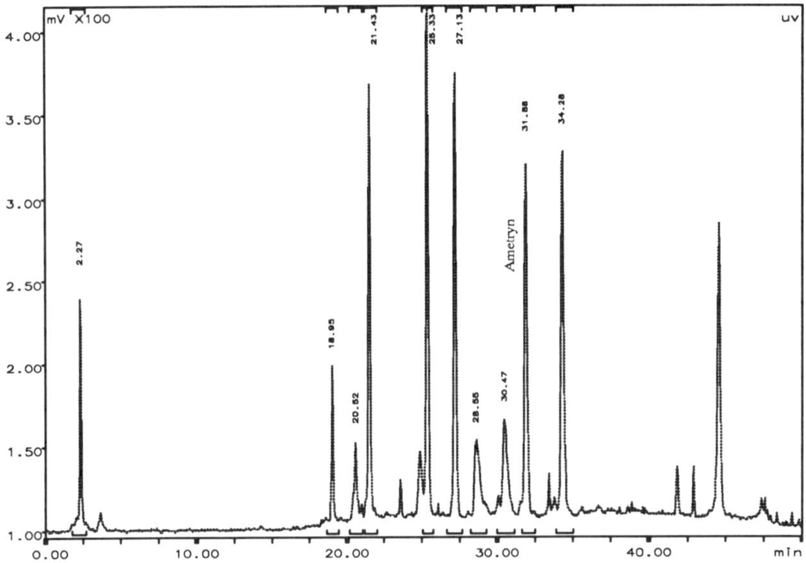

Rat Urine

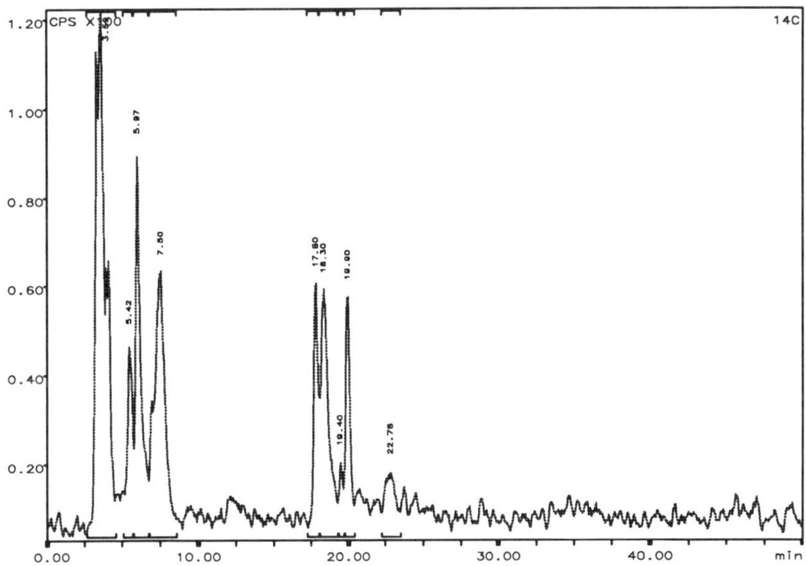

Figure 2. HPLC Chromatograms of Ametryn Reference Standards and Rat Urine

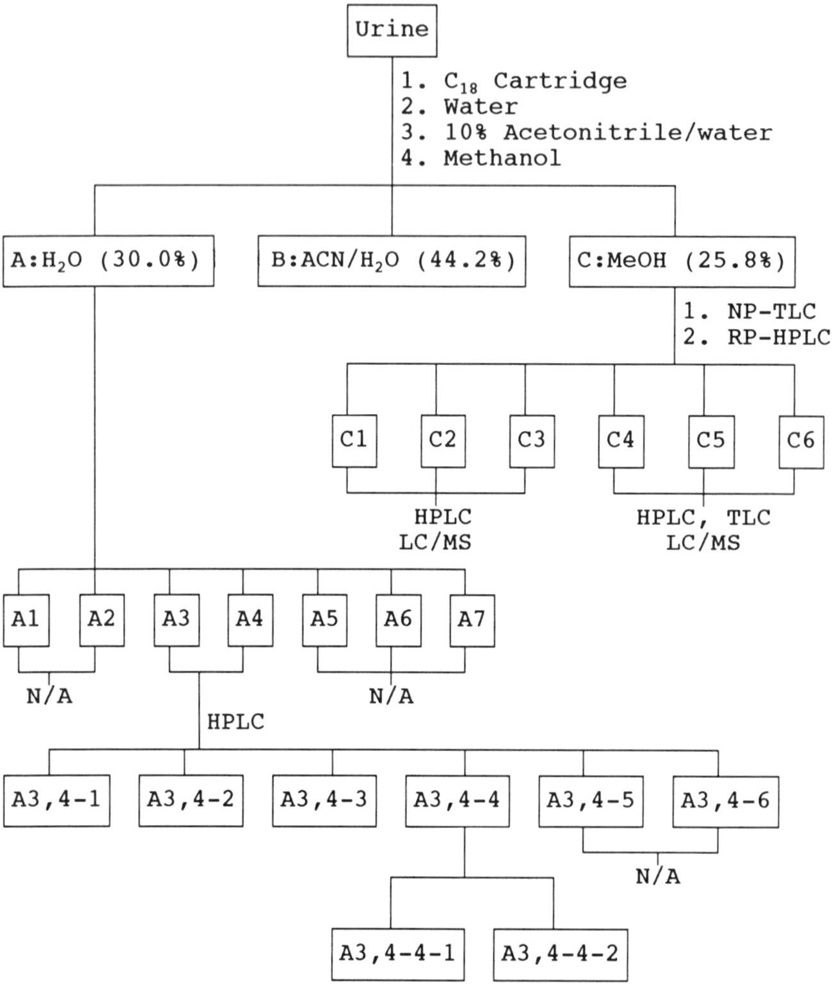

NP-TLC: Normal phase thin-layer chromatography.
RP-HPLC: Reverse phase high performance liquid chromatography.
HPLC: High performance liquid chromatography.
TLC: Thin-layer chromatography.
LC/MS: Liquid chromatography/mass spectrometry.
N/A: Not analyzed (insignificant amount).

Figure 3. Isolation Flowchart for Ametryn Rat Metabolites

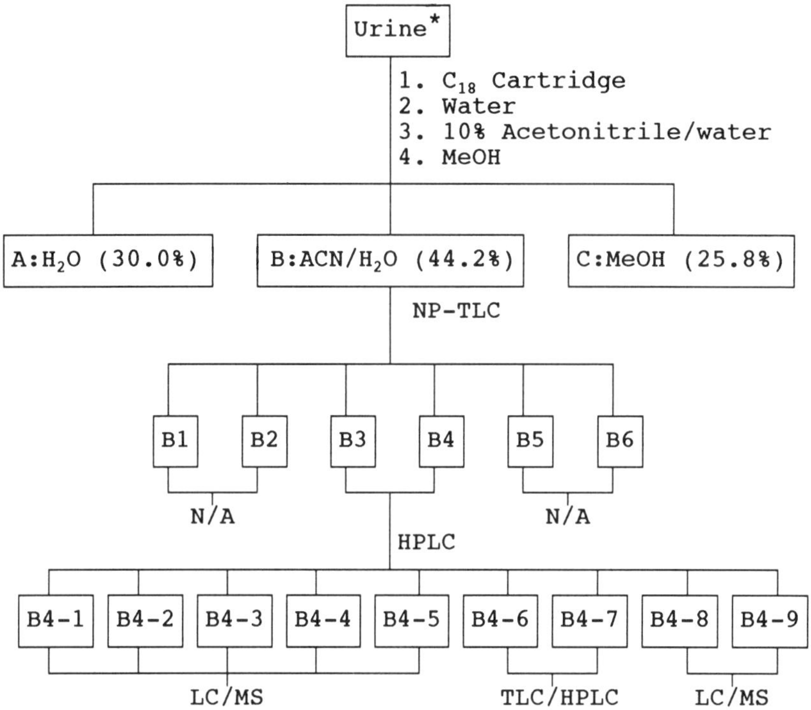

NP-TLC: Normal phase thin-layer chromatography.
N/A: Not analyzed (insignificant amount).
HPLC: High performance liquid chromatography.
LC/MS: Liquid chromatography/mass spectrometry.
TLC: Thin-layer chromatography.

*Metabolite codes RBC 6-4-6-1, RBC 6-4-7A, H-3-A, H-3'-A, H-7-1, and DD-2 were derived from additional isolation and purification procedures.

Figure 3. Isolation Flowchart for Ametryn Rat Metabolites *(Continued)*

Table IV. Mass Spectral Summary of Isolated Metabolites

Product	Fraction	Code	Chemical Name	Chemical Structure	MW (m/z)	Mass Mode
1	C-1	CG-13	Atrazine disulfide	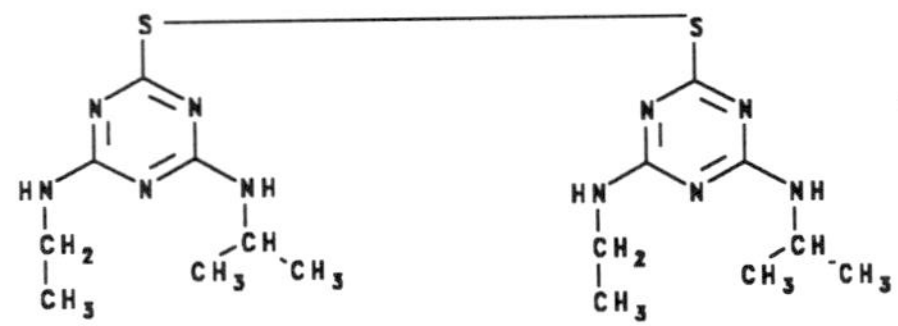	424 (425)	TSP/LC/MS
2	C-2	M1	6-Methoxy atrazine	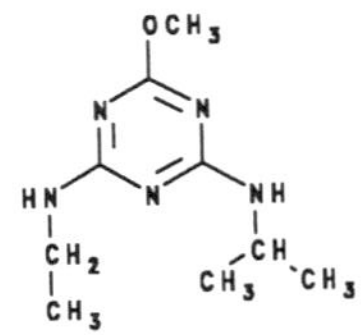	211 (212)	TSP/LC/MS
3	C-2	M2	Mono-de-ethyl atrazine disulfide	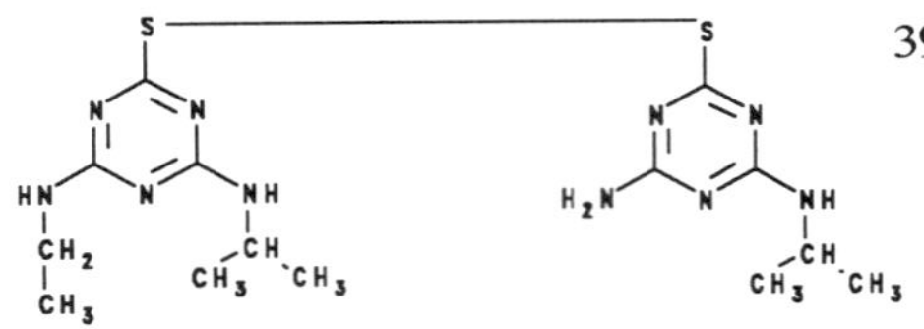	396 (397)	TSP/LS/MS

4	C-4-6	CG-11	*N*-Acetylcysteine atrazine	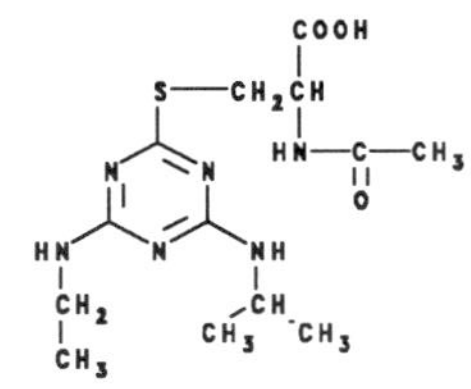	342 (343)	TSP/LC/MS
5	C-3	CG-2	Diamino ametryn	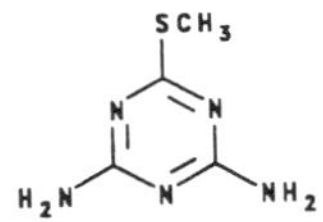	157 (158)	TSP/LC/MS
6	RBC 6-4-6-1	M3	*N*-De-ethyl-*N*-acetylcysteine atrazine	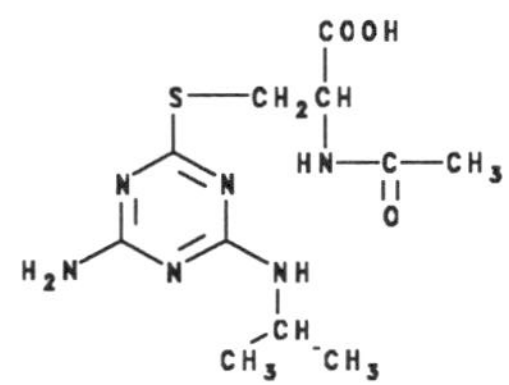	314 (315) 314 (313)	FAB/(+)/MS FAB/(-)/MS

Continued on next page

Table IV. Mass Spectral Summary of Isolated Metabolites *(Continued)*

Product	Fraction	Code	Chemical Name	Chemical Structure	MW (*m/z*)	Mass Mode
7	B-4-6,7	M10	*N*-De-isopropyl-*N*-acetyl-cysteine atrazine	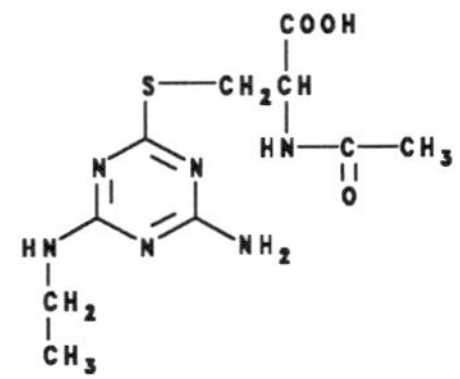	300 (301)	TSP/LC/MS
8	RBC 6-4-7A	M4	*N*-Isopropanol-*N*-acetyl-cysteine atrazine	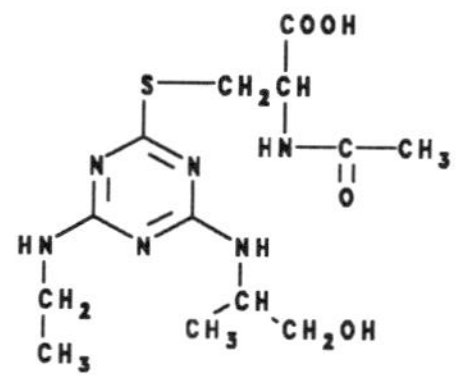	358 (359) 358 (357)	FAB/(+)/MS FAB/(-)/MS
9	H-3'-A	M5	2,4-Dihydroxy-*N*-acetylcysteine atrazine	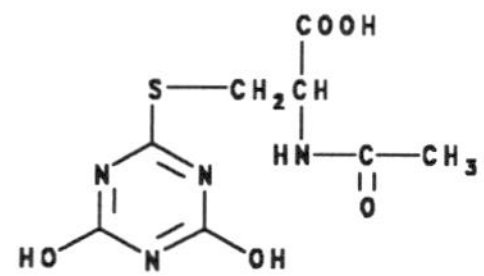	274 (275) 274 (273)	FAB/(+)/MS FAB/(-)/MS

10	H-3-A	M6	*N*-Isopropionate ametryn	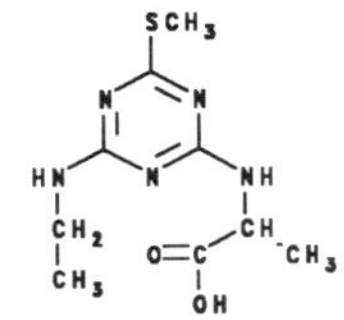	257 (258) 257 (256)	FAB/(+)/MS FAB/(-)/MS
11	DD-2	M7	*N*-De-ethyl-*N*-isopropionate ametryn	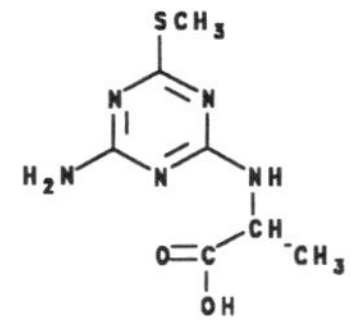	229 (230) 229 (230) 229 (228)	TSP/LC/MS FAB/(+)/MS FAB/(-)/MS
12	A-3-4-4-2	M8	2-*O*-Sulfonate ametryn	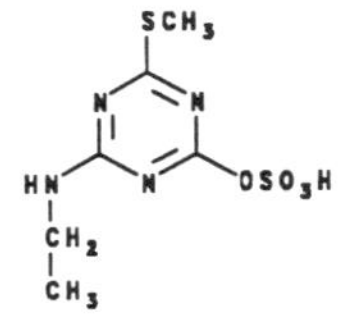	266 (265)	FAB/(-)/MS
13	H-7-1	M9	6-Thioglucuronate atrazine	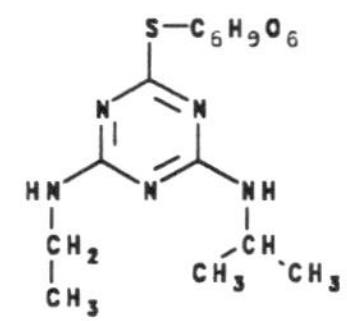	389 (390) 389 (388)	FAB/(+)/MS FAB/(-)/MS

Examples of Rat Metabolite Structure Elucidation

Several examples of the approach used to elucidate metabolite structures merit additional discussion. The mass fragmentation pattern of the first example showed a protonated molecular ion at *m/z* 343 $(M+H)^+$ (Figure 4). Other significant fragments at *m/z* 239, 214, and 197 are indicative of a substituted parent compound. A TLC and an HPLC comparison of this product with the reference metabolite standard, *N*-acetylcysteine atrazine (CG-11) verified the identity of this product.

Another product, *N*-isopropionate ametryn (Product 10), was isolated from a urine fraction designated as H-3-A. FAB/(-)/MS analysis showed mass anions at *m/z* 256 (M^-), *m/z* 278 $[(M^-Na^+)-H]^-$, *m/z* 535 $[M^-(M^-Na^+)]$, *m/z* 432 (278 + 154 from matrix), and *m/z* 557 $[(M^-Na^+)-H]^- + (M^-Na^+)$. FAB/(+)/MS analysis of this product showed protonated mass ions at *m/z* 258 $[(M^-H^+)+H]^+$, *m/z* 280 $[(M^-Na^+)+H]^+$, *m/z* 302 $[(M^-Na^+)+Na]^+$, *m/z* 412 (258 + 154 from matrix), and *m/z* 456 (302 + 154 from matrix). The molecular ion was, therefore, determined to be at *m/z* 257, which is 30 mass units higher than the parent compound. Substitution of one methyl group with a carboxyl (COOH) moiety would give a 30 mass unit increment. There are two possible locations for this structure modification, namely, at the ethyl carbon or at the isopropyl methyl carbon. Conversion of isopropyl to isopropionate appears to be more likely since a corresponding metabolite hydroxylated at the isopropyl moiety was also identified. Another related acid metabolite (discussed below) showed a loss of the *N*-ethyl group. This product was, therefore, designated as *N*-isopropionate ametryn.

The third example to be discussed is *N*-de-ethyl-*N*-isopropionate ametryn (Product 11). FAB/(-)/MS analysis of the product showed the mass anions at *m/z* 228 $(M-H)^-$ and *m/z* 250 $[(M^-Na^+)-H]^-$. FAB/(+)/MS analysis of this metabolite showed protonated mass ions at *m/z* 230 $(M+H)^+$ and *m/z* 252 $(M+Na)^+$. From the mass spectroscopic data, the molecular ion at *m/z* 229 was assigned for the metabolite, which is two mass units higher than the parent chemical and 28 mass units lower than Product 10 (last example), equating to a loss of the ethyl moiety from the molecule. Product 11 was, therefore, identified as *N*-de-ethyl-*N*-isopropionate ametryn.

Product 11 was methylated with diazomethane, followed by DIP/MS analysis. The EI mode of DIP/MS showed detailed mass fragments for the derivatized metabolite. Mass ions at *m/z* 243 (molecular ion), *m/z* 212 (M^+ - 31 or OCH_3), *m/z* 211 (M^+ - 32 or CH_3OH), and *m/z* 184 (M^+ - 59 or $COOCH_3$), all relate to a methylated carboxyl group. Mobility of Product 11 on HPLC is affected by the pH of water in the mobile phase, which is indicative of the presence of an ionic (COOH) functional group.

Summary of the Isolated and Identified Rat Metabolites. Figure 5 illustrates the proposed metabolic pathways of ametryn in the rat. Conjugation with glutathione at the 6-position, with subsequent conversion to mercapturic acid, followed by *N*-dealkylation, appeared to be one of the major metabolic pathways. Another significant metabolic pathway involved (1) dealkylation and apparent hydrolytic deamination with subsequent *O*-sulfate conjugation, and (2) *N*-isopropyl oxidation coupled with *N*-de-ethylation to form the corresponding isopropanol and isopropionate analogs. Dimerization with a disulfide linkage and formation of thioglucuronate were two minor pathways.

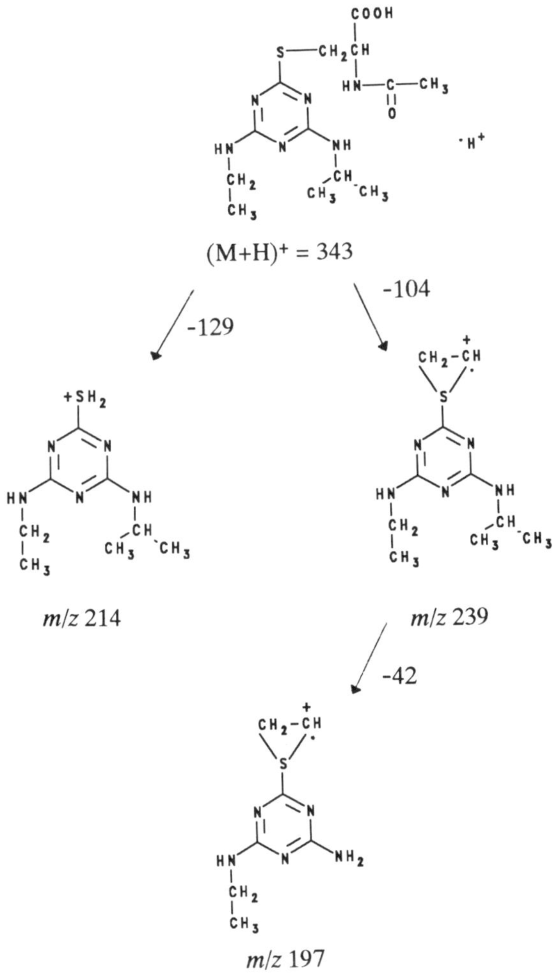

Figure 4. Mass Fragmentation Pattern of *N*-Acetylcysteine Atrazine (CG-11)

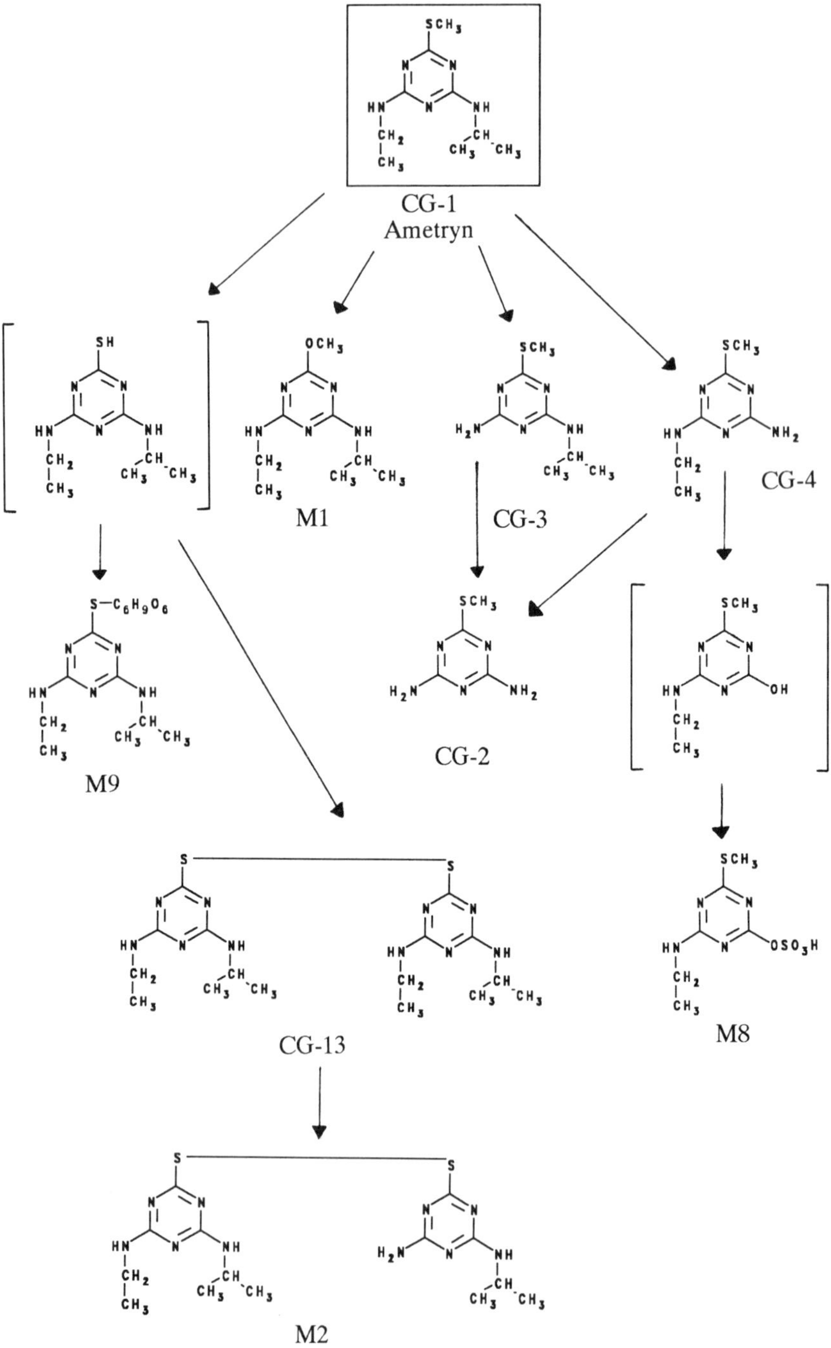

Figure 5. Proposed Pathways of Ametryn Metabolism in the Rat

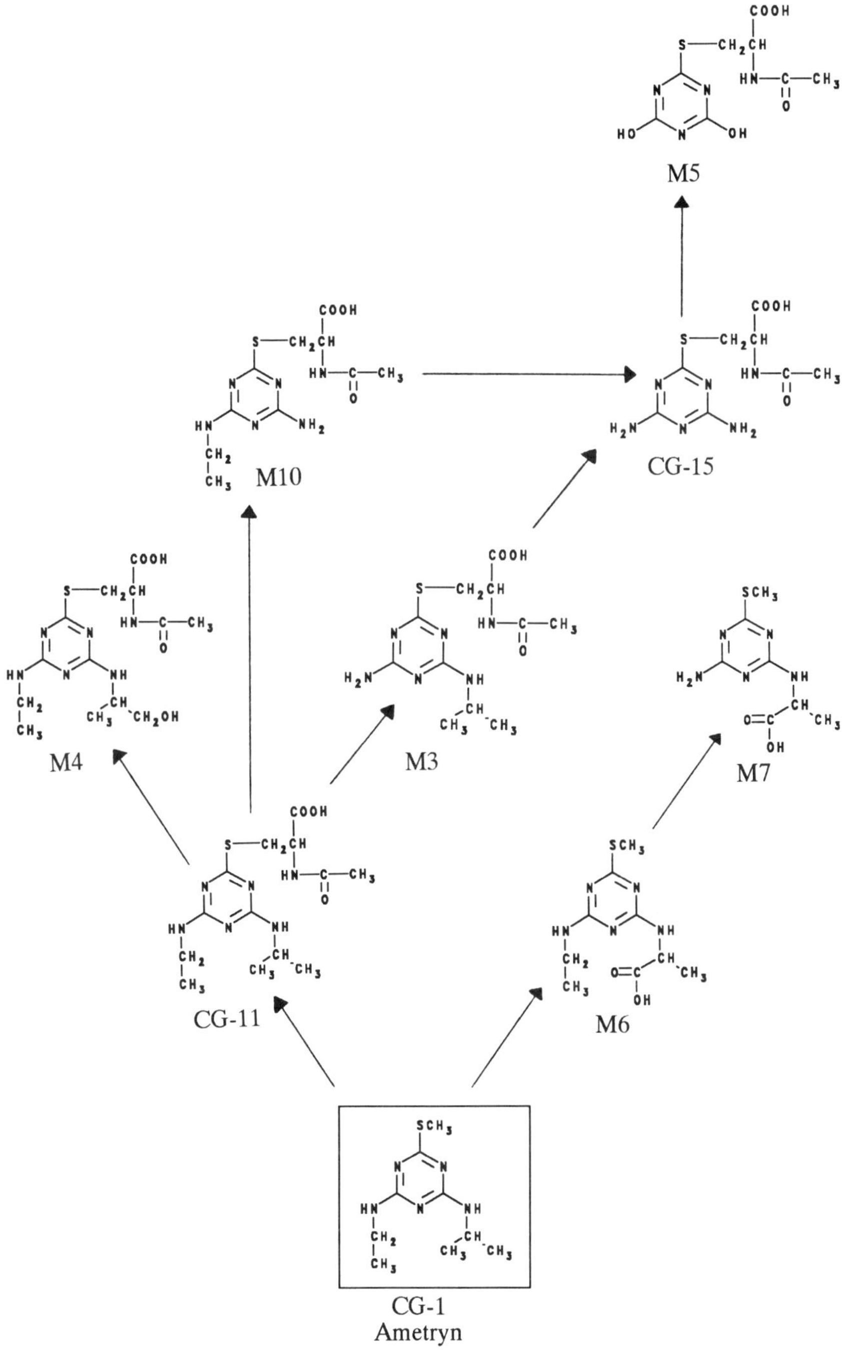

Figure 5. Proposed Pathways of Ametryn Metabolism in the Rat *(Continued)*

HPLC and TLC Analyses of Goat and Hen Excreta Metabolites

The initial strategy to conduct the livestock metabolism study involved the comparative analysis of excreta using analytical methodology, as well as synthetic reference chemicals and metabolites obtained from the rat; the overall approach, however, proved unsatisfactory. Analytical methods were applicable, but a majority of the goat and hen metabolites did not match the available reference standards. Figure 6 shows an example of an HPLC chromatogram of rat urine and hen excreta. Most of the radioactive peaks were spread over a wide range of retention times during the entire HPLC analysis.

Hydrolysis data suggested that most of the excreta radioactivity could be covalently bound to proteins. After enzyme hydrolysis, radioactive residues remained conjugated in some form with polypeptides or oligopeptides. To simplify the analysis, acid hydrolysis using both 1 N and 6 N hydrochloric acid (HCl) was employed. The 1 N HCl hydrolysis gave results similar to those obtained after enzyme hydrolysis. The 6 N HCl hydrolysis, on the other hand, which was performed under a vacuum to achieve complete deconjugation of proteins and polypeptides, gave much more simplified metabolite profiles as analyzed by HPLC. Figure 7 shows a comparison of the HPLC chromatograms of goat urine samples before and after 6 N HCl hydrolysis. It is evident that the 6 N HCl hydrolysate was much simpler to analyze. Identification of hydrolyzed aglycones was based on comparison of the HPLC R_t and TLC R_f (ratio-to-front) values with those of the reference standards, including those prepared as discussed in the following section.

Conversion of Metabolite Standards to Hydroxylated Triazines

Due to lack of adequate TLC visualization, secondary confirmation of some metabolites by TLC comparison with synthetic reference standards was not possible. Chemical conversion of a number of thiomethyl triazines to their respective hydroxy-derivatives was carried out in order to obtain more easily visualized products. After 6 N HCl hydrolysis, ametryn was converted exclusively to 6-hydroxy-atrazine (CG-7).

Similar reactions were performed on 6-substituted ametryn analogs, such as a dealkylated derivative, diamino ametryn (CG-2); *N*-de-ethyl ametryn (CG-3); and *N*-de-isopropyl ametryn (CG-4). *N*-Acetylcysteine atrazine (CG-11) and *N*-acetylcysteine diamino triazine (CG-15), the radiolabeled materials isolated from rat, goat, and hen studies, were treated in the same manner. Data suggest that the amino side chain of the triazine molecule, either as *N*-ethyl or *N*-isopropyl, was stable under acidic reflux conditions. The 6-position with either thiomethyl or another conjugated functional group was labile and subjected to hydrolysis by 6 N HCl. Thus, ametryn and its conjugated metabolites (possibly with proteins or polypeptides), as well as its corresponding *N*-dealkylated metabolites, after hydrolysis by 6 N HCl, could liberate *N*-dealkylated and hydroxylated triazine derivatives.

Extraction and Fractionation of ^{14}C Residues From the Goat and Hen

Table V shows an extraction summary of ^{14}C residues from goat and laying hen liver. Data show that the majority of the ^{14}C radioactivity was associated with the nonextractable residues (i.e., post extraction solids, PES). In general, most of the nonextractable residues could be released by protease enzyme as aqueous-soluble metabolites, which were polar. The final bound (nonextractable) residues amounted to less than 10% in all of the tissues and organs analyzed.

Rat Urine

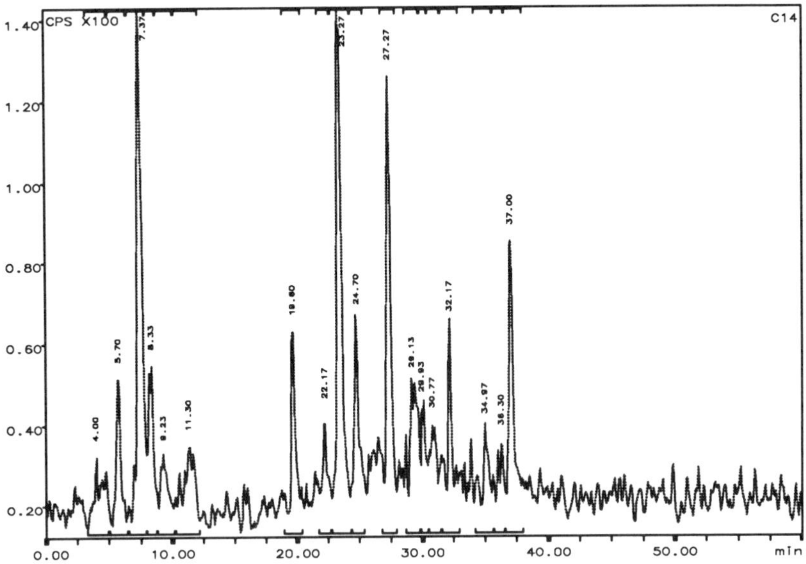

Hen Excreta MeOH/H_2O Fraction

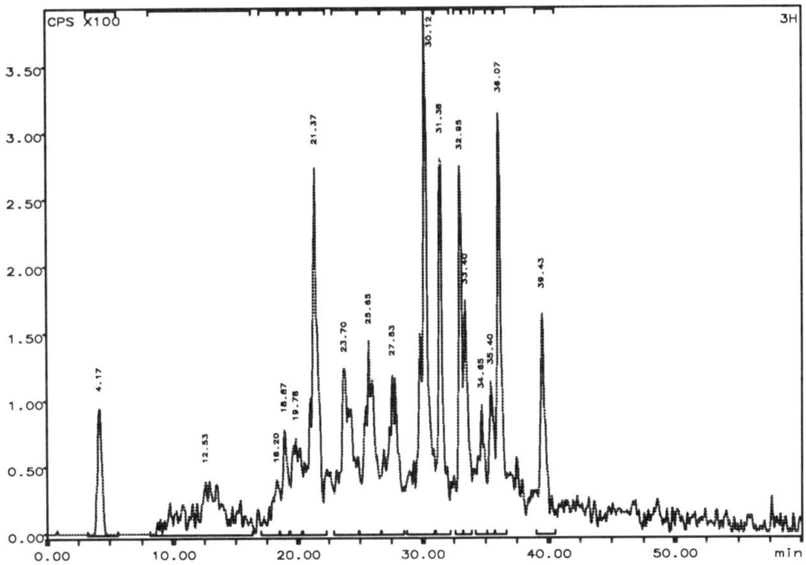

Figure 6. HPLC Chromatograms of Rat Urine and Hen Excreta

Goat Urine

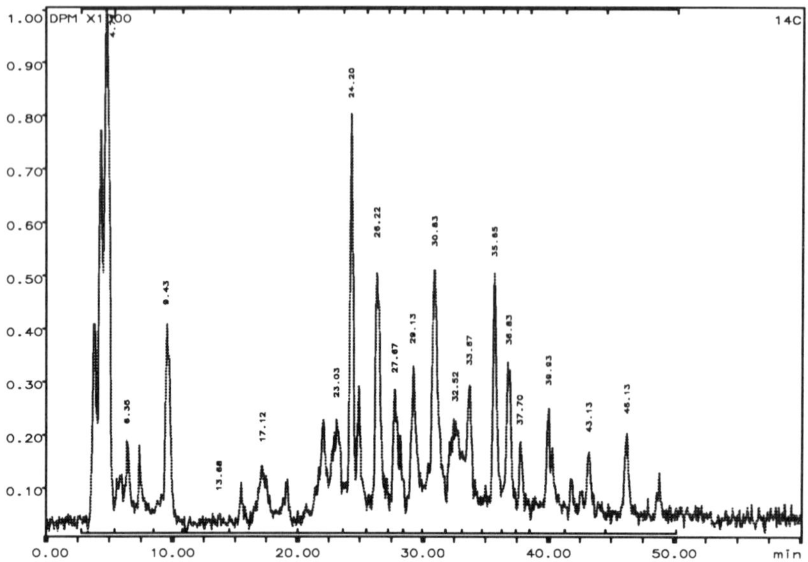

Goat Urine/6 N HCl

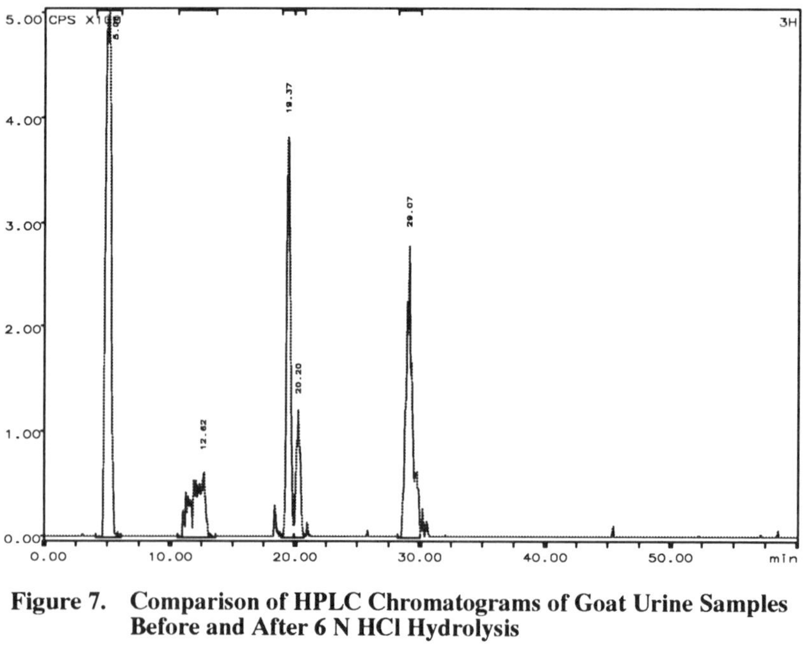

Figure 7. Comparison of HPLC Chromatograms of Goat Urine Samples Before and After 6 N HCl Hydrolysis

Table V. Extraction Profile in Ametryn-Treated Goat and Hen Liver

Fraction	*Goat*	*Hen*
$CHCl_3$	9	2
$MeOH/H_2O$	30	19
Nonextractable	61	79

HPLC analysis of the goat liver aqueous fraction or the enzyme-hydrolyzed aqueous fraction yielded a complex profile. Attempts at confirmation by TLC yielded inconclusive results. Preparative purification by HPLC, followed by silica gel TLC, indicated that the majority of the metabolites were not identical to those obtained from the rat. This is probably due to the oligopeptides still being associated with the triazine moiety. Similar observations were found in the laying hens. The polar metabolites in the laying hen, either in the $MeOH/H_2O$ or in the enzyme-hydrolyzed aqueous fractions, showed behaviors similar to those in the goat. However, acid digestion of the PES by 6 N HCl resulted in a very simplified product spectrum (Figure 8). The products identified by this technique included the following: 2-amino-4-isopropylamino-6-hydroxytriazine (CG-5), CG-7, 2-amino-4,6-dihydroxytriazine (CG-9), 2,4-diamino-6-hydroxytriazine (CG-16), and 2-amino-4-ethylamino-6-hydroxytriazine (CG-17). The technique was also successfully employed on kidney, liver, and muscle PES of both animals. Proposed pathways of ametryn metabolism in the lactating goat and the laying hen are illustrated in Figure 9.

Summary and Conclusions

Comparing the results obtained from the rats, lactating goats, and laying hens, it appears that the common, free, nonconjugated metabolites included the following: *N*-de-isopropyl ametryn, *N*-de-ethyl ametryn, diamino ametryn, and a trace of parent compound. In the various tissue matrices of the goats and hens, the products included conjugated metabolites, all containing an intact triazine ring associated (or conjugated) with endogenous components, such as proteins or polypeptides. In reviewing the metabolic pathways in the rats, goats, and hens, it can be concluded that they appear to share similar metabolic pathways.

In conclusion, ametryn was readily absorbed, extensively metabolized, and then excreted after it was administered to the rats, lactating goats, and laying hens. The majority of the dosed chemical was eliminated in excreta. In addition to the parent compound ametryn, a number of other metabolites including polar and nonpolar, conjugated and nonconjugated metabolites were detected. *N*-Dealkylation of the molecule and conjugation with peptides were the major routes of biotransformation in the goats and hens. *N*-Dealkylation, oxidation/hydroxylation, and conjugation (with sulfate, glutathione, and glucuronic acid) were predominant in the rats. Hydrolysis of the conjugated metabolites with a strong acid to hydroxytriazine showed that the triazine ring was intact in these metabolites. This procedure also effects a very considerable simplification of the tissue metabolite profile on which routine analytical methods are based.

Goat Liver PES/6 N HCl

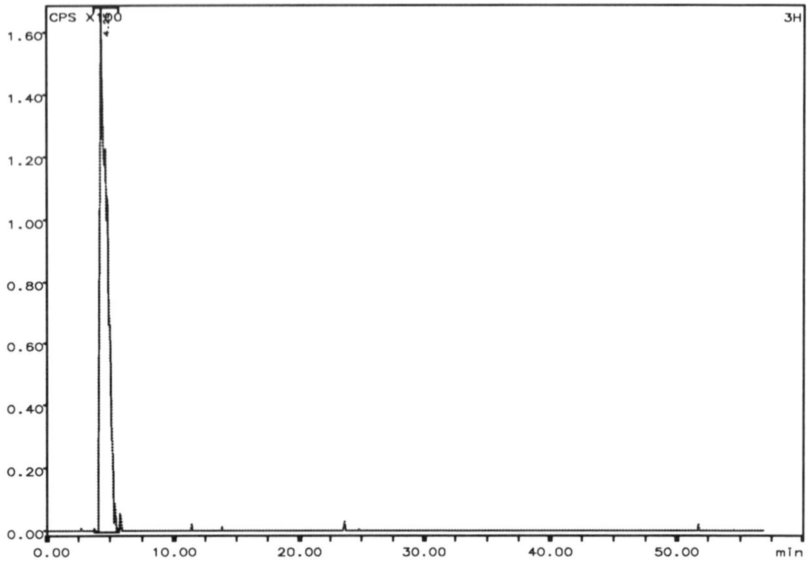

Hen Liver PES/6 N HCl

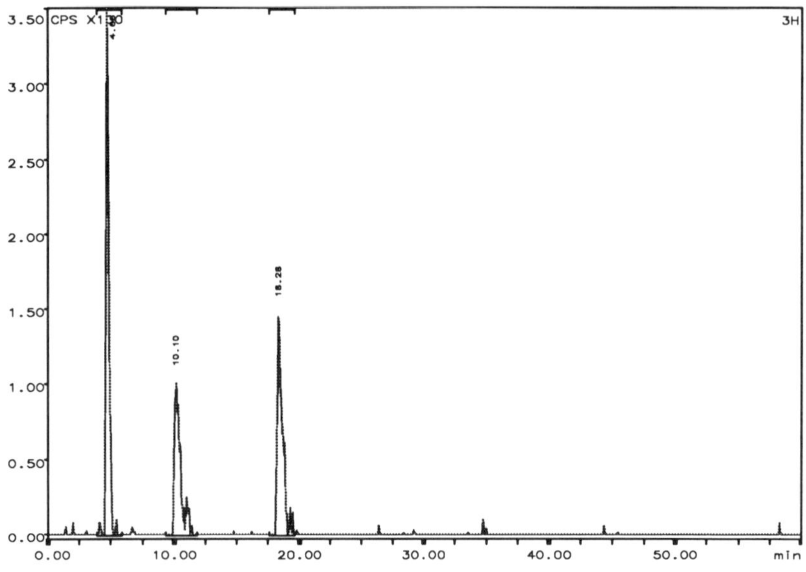

Figure 8. HPLC Chromatograms of Goat and Hen Liver PES After 6 N HCl Hydrolysis

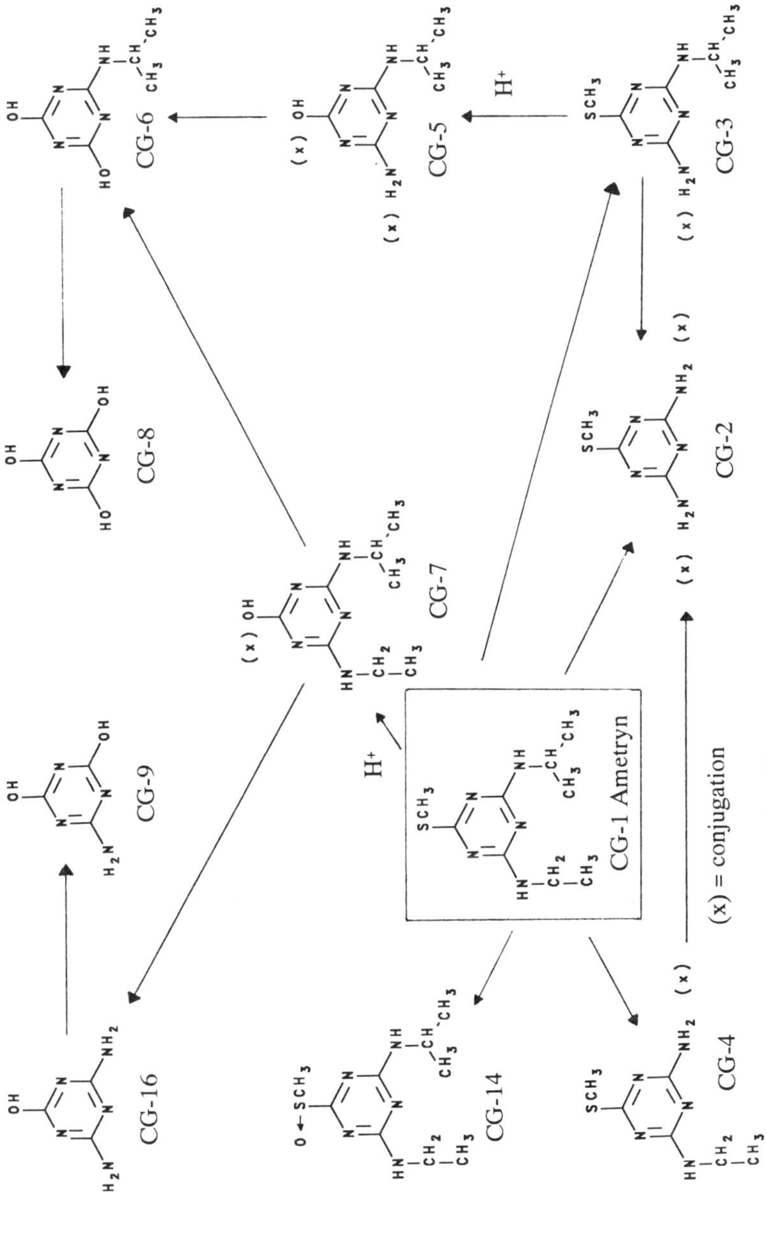

Figure 9. Proposed Pathways of Ametryn Metabolism in the Lactating Goat and Laying Hen

RECEIVED April 13, 1992

Chapter 13

Sulfonamide Drug in Lactating Dairy Cows

Novel Metabolism

G. D. Paulson, V. J. Feil, P. J. Sommer, and C. H. Lamoureux

Agricultural Research Service, Biosciences Research Laboratory, U.S. Department of Agriculture, P.O. Box 5674–University Station, Fargo, ND 58105–5674

Lactating dairy cows were orally or intravenously dosed with ^{14}C-Sulfamethazine {^{14}C-Sulmet; 4-amino-N-(4,6-dimethyl-2-pyrimidinyl)-benzene[U-^{14}C] sulfonamide}. ^{14}C-Labeled compounds in tissues and urine were isolated by chromatographic techniques and identified by ^{1}H-NMR and mass spectrometry. The cows metabolized ^{14}C-Sulmet by the following mechanisms: (1) acetylation at the N^4-position; (2) hydroxylation at the 5-position of the heterocyclic ring followed by sulfate ester or hexuronic acid conjugation; (3) oxidation of a methyl group to a hydroxymethyl group followed by sulfate ester conjugation; (4) conjugation of the N^1-position with a hexose or a hexuronic acid; (5) hydroxylation of the 3-position of the benzene ring followed by hexuronic acid conjugation; and (6) cleavage of the N^1-C bond to yield sulfanilamide.

Littlefield et al. (*1*) reported that the antibiotic sulfamethazine [4-amino-N-(4,6-dimethyl-2-pyrimidinyl)benzenesulfonamide] induced hyperplasia and adenomas of the thyroid glands in rats. This report and the results of surveys indicating that sulfamethazine residues were present in a large percentage of market milk samples collected in the USA and Canada (*2-4*) have generated concern about the disposition of sulfamethazine in dairy animals. The studies reported here were initiated to compare the metabolic fate of ^{14}C-sulfamethazine {4-amino-N-(4,6-dimethyl-2-pyrimidinyl)benzene-[U^{14}C]sulfonamide} when given orally or intravenously to lactating dairy cows.

Materials and Methods

Chemicals. ^{14}C-Sulfamethazine (obtained from Midwest Research Institute, Kansas City, MO) was purified by reverse phase high pressure liquid chromatography. The HPLC conditions were as follows: 8 mm x 10 cm NOVA-PAK

Published 1992 American Chemical Society

C-18 Cartridge (4 μm particle size) in a Radial Compression Cartridge Holder; solvent, 75% H_2O - 25% methanol. Ninety-nine percent of the ^{14}C activity had the same retention time as reference sulfamethazine when chromatographed using HPLC-1 conditions as described below. The synthesis and purification of N^4-acetylsulfamethazine has been reported (*5*).

Dosing of Animals and Collection of Samples. Lactating holstein cows (570-618 kg; 4-5 years old; second month of lactation; milk production, 35-37 kg/day) used in these studies were obtained from a commercial dairy farm. The animals were fed a ration designed for cows in heavy lactation (9 kg grain concentrate per day; high quality alfalfa hay, trace mineralized salt and water ad libitum) prior to and after dosing. One cow was orally dosed with ^{14}C-sulfamethazine (220 mg/kg body weight; specific radioactivity (SA) 100 dpm/μg sulfamethazine). The dose was placed in gelatin capsules and administered with a balling gun. Another cow was intravenously dosed with ^{14}C-sulfamethazine (220 mg/kg body weight; SA 100 dpm/μg of sulfamethazine; dose dissolved in approximately 800 ml of H_2O; adjusted to pH 10.5 with NaOH; dose perfused over a 30 minute period) via an indwelling catheter in the jugular vein.

The cows were maintained in box stalls and blood (approximately 20 ml at each collection time via indwelling jugular catheters 1, 3, 6, 12, 24, 36, and 48 hours after dosing), milk (0-12, 12-24, 24-36, and 36-48 hours after dosing), and tissues (48 hours after dosing) were collected and stored at -15°C until analyzed.

Isolation and Identification of ^{14}C-Labeled Fractions. The sequences of reverse phase chromatographic procedures used to isolate ^{14}C-labeled compounds from cow urine are outlined in Figure 1. Waters Associates column systems [8 mm x 10 cm Radial Compression Cartridge Holder, NOVA-PAK C18 (4 μm particle size) Cartridge] and a flow rate of 3 ml/minute were used for all chromatographic procedures. Other conditions were as follows: HPLC-1, 30 minute gradient (Waters Associates gradient 7) from 95% 1 mM phosphate buffer, pH 7.0 and 5% methanol to 100% methanol; HPLC-2, two columns in series, 7 minute isocratic 0.1N acetic acid followed by 20 minute gradient (Waters Associates gradient 7) to 50% 0.1N acetic acid and 50% methanol; HPLC-3, two columns in series, 30 minute gradient (Waters Associates gradient 7) from 95% 0.1N acetic acid and 5% methanol to 100% methanol; HPLC-4, two columns in series, isocratic 0.1N acetic acid; HPLC-5, two columns in series, 30 minute gradient (Waters Associates gradient 8) from 95% 0.1N acetic acid and 5% methanol to 100% methanol; HPLC-6, two columns in series, 30 minute gradient (Waters Associates gradient 7) from 95% water and 5% methanol to 100% methanol; HPLC-7, two columns in series, 30 minute gradient (Waters Associates gradient 6) from 95% 0.1N acetic acid and 5% methanol to 100% methanol; HPLC-8, 30 minute gradient (Waters Associates gradient 8) from 95% 1 mM phosphate buffer, pH 7.0 and 5% methanol to 100% methanol; HPLC-9, isocratic 88% 1 mM phosphate buffer, pH 7.0 and 12% methanol. The purified ^{14}C-labeled fractions (Figure 1) were characterized by ^1H-NMR and/or mass spectrometry and structures were assigned as outlined in Table IV.

Tissue samples were analyzed for ^{14}C-labeled sulfamethazine, N^4-acetylsulfamethazine and other fractions using a described methodology (*6*). In brief, this involved spiking homogenized tissues with non-radiolabeled reference compounds, methanol extraction, hexane-water partitioning followed by reverse phase chromatography (XAD-2 and C-18 HPLC). The ^{14}C activity in each fraction was quantified by liquid scintillation counting.

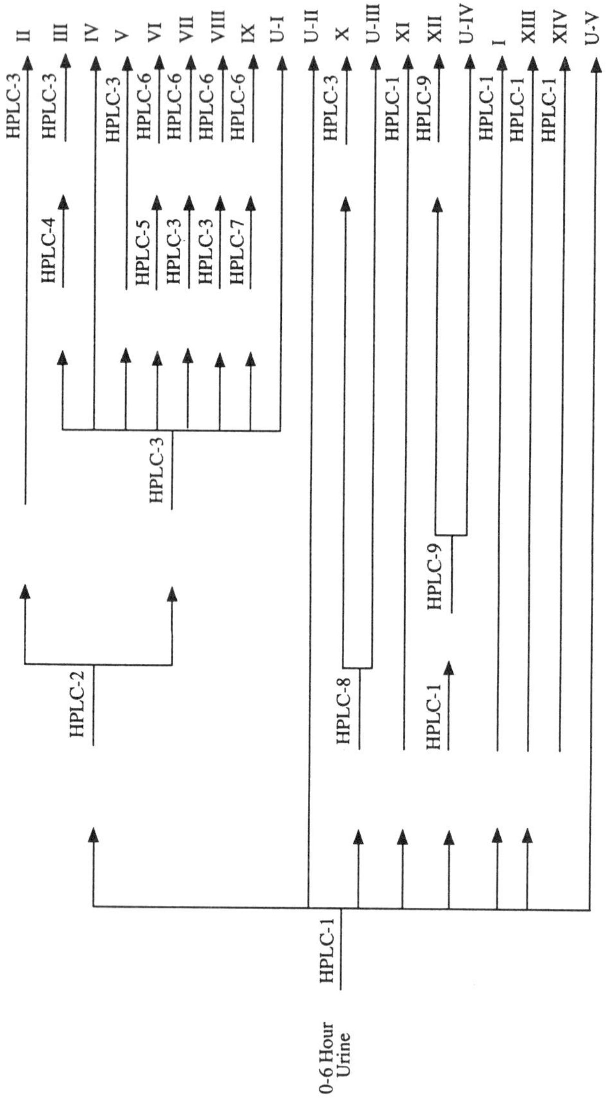

Figure 1. Isolation of ^{14}C-fractions from urine of cows dosed with ^{14}Csulfamethazine.

Instrumentation. Tissue and feces (approximately 500 mg/assay) were combusted with a Model 307 Packard Oxidizer and the resulting $^{14}CO_2$ was trapped in Carbosorb, (Packard Instrument Co., Downers Grove, IL) The ^{14}C-activity was measured by liquid scintillation counting (Model 1600 CA Packard liquid scintillation counter). Milk and urine (approximately 500 mg/assay) were mixed with 15 ml of Ecolite (ICN Biomedicals Inc., Irvine, CA) and the ^{14}C-activity was measured with a Model LS 1701 Beckman liquid scintillation counter. Ecolite and liquid scintillation counting was also used to measure ^{14}C-activity in chromatographic fractions (2-6 ml). Fast atom bombardment (FAB) mass spectra and electron impact (EI) mass spectra were obtained with a Varian model CH-5 DF mass spectrometer interfaced to a SS-200 data system. The beam of fast atoms was provided with an ION TECH Saddle-field gun operated at 8KV with xenon gas. Glycerol, or glycerol + Proton Sponge (Aldrich Chemical Co.) were used as matrices for FAB analyses. A Bruker AM 400 spectrometer was used to obtain nuclear magnetic resonance (NMR) spectra (using TMS as the reference standard). A Waters Associates, Inc. System (two model 6000A pumps; model 440 UV detector with a 254 filter; model 680 gradient controller; model 746 data module; and a RCM 8 X 10 cm Radial Compression Cartridge Holder) was used for high pressure liquid chromatography. Injections were made with a model 7125 Rheodyne injector.

Results and Discussion

The concentration of ^{14}C-activity in the blood of the orally dosed cow increased from 1 to 24 hours after dosing with ^{14}C-sulfamethazine and then declined. The ^{14}C-residues in the blood of the intravenously dosed cow rapidly declined to approximately 1/5 the level observed in the blood of the orally dosed cow 48 hours after dosing (Table I). ^{14}C-Activity was widely distributed throughout the bodies of both cows 48 hours after dosing (Table I) and most of the tissues from the orally dosed cow contained 2-4 times more ^{14}C-activity than the tissues from the intravenously dosed cow. One notable exception was that the skin from the intravenously dosed animal contained approximately four times more ^{14}C-residue than found in the skin of the orally dosed cow. We do not have an explanation for this observation; however, it is interesting that sulfadimethoxine was eliminated more slowly from the skin than from other tissues or plasma when this compound was administered to chickens (*7*).

The urine collected from 0-48 hours after dosing accounted for 54.3% and 73.3% of the ^{14}C-activity given orally and intravenously, respectively. The rate of excretion of ^{14}C-activity in the urine was highest from 0-24 hours after dosing and then declined (Table II). These results are in general agreement with earlier observations on urinary excretion of sulfamethazine related residues in animal urine (*5, 8-12*).

Fecal excretion (0-48 hours after dosing) accounted for 16.8% of the orally administered ^{14}C-activity and 7.4% of the intravenously administered ^{14}C-activity (Table II). Forty-eight hours after intravenous dosing the ^{14}C concentration was higher in the upper part of the gastrointestinal tract (rumen, reticulum, omasum and abomasum) than in the intestines as shown in Table I. Thus the ^{14}C-activity in the GIT and the feces from the intravenously dosed cow could not have resulted entirely from biliary secretion. Salivary secretion of sulfonamide drugs, including sulfamethazine, has been reported (*13-16*) and may have been at least partially responsible for the ^{14}C-activity entering the GIT of the intravenously dosed cow.

Table I. Concentration of Total ^{14}C-Residues (Sulfamethazine Equivalents) in Tissues of Cows Dosed Orally or Intravenously with ^{14}C-Sulfamethazine[a]

Tissue	Hours After Dosing	Route of Administration	
		Oral (ppm)	Intravenous (ppm)
Blood	1	72	391
Blood	3	131	309
Blood	6	171	245
Blood	12	189	166
Blood	24	140	83
Blood	36	90	36
Blood	48	52	11
Rumen, Reticulum, Omasum, Abomasum (including contents)	48	106	37
Intestines (including contents)	48	106	26
Kidney	48	71	20
Liver	48	51	16
Lung	48	33	7
Heart	48	29	5
Spleen	48	27	4
Skeletal Muscle	48	23	4
Udder	48	21	6
Adipose	48	12	3
Skin	48	12	49
Bone	48	7	7

[a] Cows were dosed with ^{14}C-Sulfamethazine (SA = 100 dpm/µg; 220 mg of sulfamethazine per kg of body weight).

Table II. Disposition of ^{14}C-Activity Administered to Cows as ^{14}C-Sulfamethazine[a]

Fraction	Route of Administration	
	Oral	Intravenous
	(percent of dose)	
0-12 Hour Urine	14.6	39.2
12-24 Hour Urine	20.3	17.4
24-36 Hour Urine	9.8	11.2
36-48 Hour Urine	9.6	5.5
0-12 Hour Feces	0.4	0.4
12-24 Hour Feces	6.3	0.5
24-36 Hour Feces	4.8	4.5
36-48 Hour Feces	5.3	2.0
0-48 Hour Milk	2.0	1.1
Tissues 48 Hours After Dosing	17.6	5.5

[a] Cows were dosed with ^{14}C-Sulfamethazine (SA 100 dpm/μg; 220 mg/kg of body weight). Urine was collected from 0-48 hours after dosing. The cows were sacrificed 48 hours after dosing and tissues were collected.

The milk collected from 0-48 hours after dosing accounted for 2.0% and 1.1% of the ^{14}C-activity administered orally and intravenously, respectively (Table II). The presence of ^{14}C-labeled sulfamethazine, the N^4-lactose conjugate of sulfamethazine and N^4-acetylsulfamethazine in the milk from both the orally and intravenously dosed cow has been described (Paulson et al. accepted for publication, Xenobiotica, 1992).

The overall recovery of ^{14}C-activity administered was 87.3% (intravenous dose) and 90.7% (oral dose). It seems likely that the less than complete recovery was primarily due to incomplete collection of urine and feces excreted by the animals.

Characterization studies summarized in Table III indicated that a small amount of N^4-acetylsulfamethazine was present in all of the tissues 48 hours after dosing. ^{14}C-Sulfamethazine was the major radiolabeled residue in the blood, skeletal muscle and adipose tissue from both the orally and intravenously dosed cows. In contrast, an unidentified fraction, more polar than sulfamethazine, accounted for the largest percentage of ^{14}C-activity in the liver and kidney. The latter fractions appeared to be composed of more than one ^{14}C-labeled compound, and it seems likely that they were polar compounds similar to those isolated from urine. It was also interesting that both the liver and kidney contained substantial amounts of insoluble ^{14}C-activity (not extracted by CH_3OH) and, in all of the tissues examined, more than 5% of the ^{14}C-activity was less polar than N^4-acetylsulfamethazine. We specifically examined the tissues for desaminosulfamethazine [a non-polar metabolite of sulfamethazine

Table III. ^{14}C-Labeled Fractions in the Tissues of Cows 48 Hours After Oral or Intravenous Dosing With ^{14}C-Sulfamethazine [a]

	Route of Administration									
	Oral					Intravenous				
^{14}C-Labeled Fraction From Tissue	Blood	Liver	Kidney	Skeletal Muscle	Adipose Tissue	Blood	Liver	Kidney	Skeletal Muscle	Adipose Tissue
	percent[b]					percent				
Insoluble Residue [c] After CH_3OH Extraction	3.1	13.8	4.3	14.8	1.3	5.4	29.3	8.0	29.0	2.1
Unidentified Fraction(s)[d] More Polar than Sulfamethazine	16.8	42.4	57.4	38.1	8.9	26.0	36.3	53.9	24.2	18.4
Sulfamethazine	72.6	28.5	27.8	40.0	66.3	47.2	27.3	19.3	25.6	41.5
N^4-Acetyl Sulfamethazine	2.3	2.3	9.8	1.8	2.1	1.9	2.3	6.6	2.0	2.7
Unidentified Fraction(s) Less Polar than N^4-Acetyl Sulfamethazine[e]	8.9	8.5	5.2	6.4	8.4	11.3	7.8	12.4	18.0	10.3

[a]Tissues were examined for ^{14}C-labeled fractions as previously described.
[b]All values are expressed as a percentage of total ^{14}C-activity in tissues.
[c]Residue remaining after methanol extraction; assay by combustion analysis.
[d]Includes all ^{14}C-activity eluted from XAD-2 column by H_2O and ^{14}C-activity eluted from HPLC (C-18 column) prior to sulfamethazine.
[e]Includes ^{14}C-activity in the hexane phase after hexane: water partition and all ^{14}C-activity eluted from HPLC (C-18 column) after N^4-acetylsulfamethazine.

previously observed in swine and rat tissues (*5*)]; however, we found no evidence for this compound in tissues from either the orally or intravenously dosed cow.

The ^{14}C-activity in the urine was resolved into 19 fractions by reverse phase HPLC (Figure 1). Fourteen of these fractions were characterized by ^{1}H-NMR and mass spectral analysis. Supporting information and the assigned structures are summarized in Table IV. Sulfamethazine and N^4-acetylsulfamethazine (fractions I and XIV) were the most abundant ^{14}C-labeled compounds in the urine of both cows (Table IV). Two minor metabolites in the urine from both cows were the result of direct conjugation of sulfamethazine (fractions X and XIII). It is likely the conjugating moieties were glucuronic acid and glucose; however, the information obtained in these studies did not rule out the possibility that the conjugating moieties were another hexuronic acid or hexose. We did establish that fractions X and XIII were not sulfamethazine conjugated with glucose or glucuronic acid at the N^4-position by comparing their HPLC retention times with reference compounds (*5*). Therefore, we concluded fractions X and XIII were formed by conjugation at the N^1-position. One other metabolite (fraction IX) apparently resulted from direct conjugation of sulfamethazine, presumably at either the N^1-position or the N^4-position. During the initial chromatographic procedures this fraction was much more polar than sulfamethazine (Figure 1); however, during subsequent steps in the isolation procedure, it apparently hydrolyzed to yield sulfamethazine, the compound identified by ^{1}H-NMR and mass spectrometry. Direct conjugation at the N^1- and N^4-position of sulfamethazine in animal systems was reported earlier (*5, 8-12*).

Oxidation of a methyl group to a hydroxymethyl group was involved in the formation of three urinary metabolites. Part of this primary product was conjugated with sulfate (fraction III) and part was acetylated at the N^4-position (fraction XI). Fraction VIII was very polar when initially chromatographed (eluted very early in HPLC-1 system) but it apparently hydrolyzed to yield the much less polar N^4-acetyl-4-hydroxymethyl-sulfamethazine, characterized by ^{1}H-NMR and mass spectrometry. Therefore, it is likely that fraction VIII, as secreted in the urine, was conjugated with glucuronic acid or some other polar moiety. Metabolism of sulfamethazine to 4-hydroxymethylsulfamethazine in animal systems has been reported (*8-12, 17, 18*), and Nielsen (*17*) reported circumstantial evidence (enzymatic hydrolysis to yield 4-hydroxymethylsulfamethazine) for sulfate and glucuronic acid conjugation of this metabolite.

Four urinary metabolites resulted from hydroxylation of the 5-position on the heterocyclic ring followed by conjugation at two different sites. The 5-hydroxy group was conjugated with either a hexuronic acid (fraction IV) or sulfate (fraction VI) and the N^4-position was acetylated (fractions VIII and XII). Evidence for the conversion of sulfamethazine to 5-hydroxysulfamethazine in rabbits (*18*) and cows (*17*) has been reported. Nielsen (*17*) found circumstantial evidence that 5-hydroxysulfamethazine was conjugated with sulfate in cows but reported no evidence for glucuronic acid conjugation of this compound.

Fraction V was unique in that it resulted from hydroxylation of the 3-position of the aromatic ring followed by hexuronic conjugation. Hydroxylation at the 3-position was assigned on the basis of ^{1}H-NMR absorptions at δ 7.70 (d, J=2.0 Hz) and at δ 7.57 (dd, J=8.8 Hz, J=2.0 Hz) which indicated that the two positions ortho to the sulfonamide moiety were occupied by protons. The ^{1}H-NMR absorption at δ 6.71 (d, J=8.8 Hz) provided evidence for the proton at the 5-position of the aromatic ring. The ion observed at m/z 469 (-FAB-MS, M-H) supported the structural assignment. These assignments are also compatable with the assignments for sulfamethazine and N^4-acetylsulfa-

Table IV. Evidence for structure of ^{14}C-labeled fractions in urine from cows after oral or intravenous dosing with ^{14}C-sulfamethazine

Assigned Structure & Chromatographic Fraction	Evidence for Structure	Route of Dosing: Oral (--% of ^{14}C in urine--)	Route of Dosing: Intravenous (--% of ^{14}C in urine--)
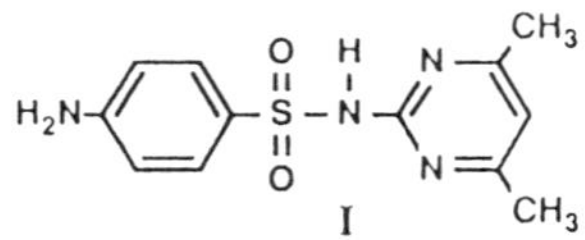 I	+FAB-MS, m/z 279 (M+H); ^{1}H-NMR (CD_3OD), δ 2.29 (s, pyrimidine-CH_3), δ 6.63 (d, J=8.8 Hz, 3,5-Ar-H), δ 6.69 (s, 5-pyrimidine-H), δ 7.74 (d, J=8.8 Hz, 2,6-Ar-H); cochromatography with reference compound	37.8	52.9
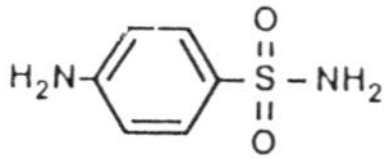 II	+FAB-MS, m/z 173 (M+H); EI-MS, m/z 172, 156, 108, 92; ^{1}H-NMR (CD_3OD), δ 6.67 (d, J=8.8 Hz, 3,5-Ar-H), δ 7.57 (d, J=8.8 Hz, 2,6-Ar-H); cochromatography with reference compound	ND	0.3
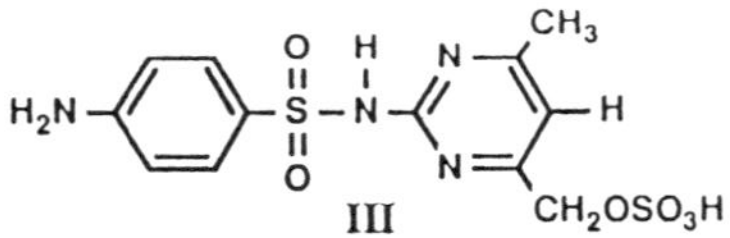 III	-FAB-MS, m/z 373 (M-H); ^{1}H-NMR (CD_3OD), δ 2.36 (s, pyrimidine-CH_3), δ 4.86 (s, pyrimidine-CH_2), δ 6.63 (d, J=8.8 Hz, 3,5-Ar-H), δ 7.02 (s, 5-pyrimidine-H), δ 7.74 (d, J=8.8 Hz, 2,6-Ar-H)	0.5	ND
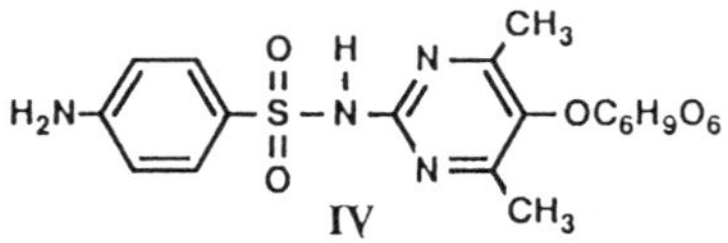 IV	-FAB-MS, m/z 469 (M-H); ^{1}H-NMR (CD_3OD), δ 2.38 (s, pyrimidine-CH_3), δ 3.39-3.71 (complex multiplet, H on C_2-C_5 of uronic acid), δ 4.46 (d, J=7.2 Hz, anomeric-H), δ 6.63 (d, J=8.8 Hz, 3,5-Ar-H), δ 7.72 (d, J=8.8 Hz, 2,6-Ar-H)	ND	1.4
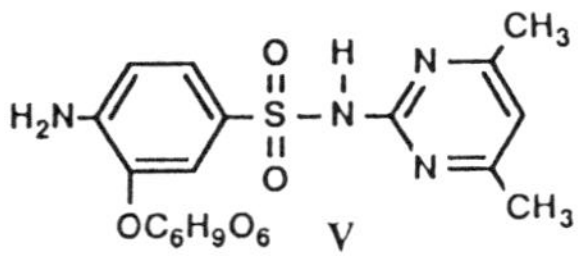 V	-FAB-MS, m/z 469 (M-H); ^{1}H-NMR (CD_3OD), δ 2.21 (s, pyrimidine-CH_3), δ 3.48-3.66 (complex multiplet, H on C_2-C_5 of uronic acid), δ 4.71 (d, J=7.2 Hz, anomeric-H), δ 6.44 (s, 5-pyrimidine-H), δ 6.71 (d, J=8.8 Hz, 5-Ar-H), δ 7.57 (dd, J=8.8 Hz, J=2.0 Hz, 6-Ar-H), δ 7.70 (d, J=2.0 Hz, 2-Ar-H)	1.2	0.8

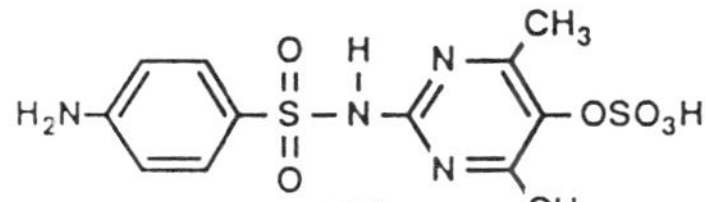 VI	-FAB-MS, m/z 373 (M-H); ^{1}H-NMR (CD_3OD), δ 2.38 (s, pyrimidine-CH_3), δ 6.62 (d, J=9.2 Hz, 3,5-Ar-H), δ 7.72 (d, J=9.2 Hz, 2,6-Ar-H)	0.2	0.6
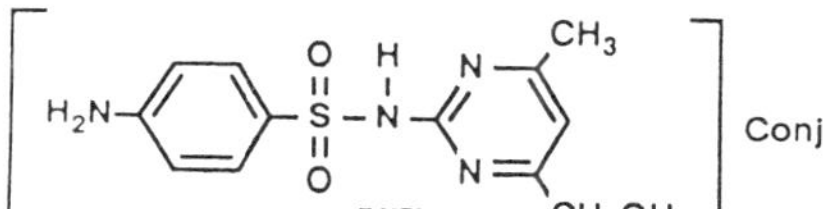 VII	Analysis of hydrolysis product; ^{1}H-NMR (CD_3OD), δ 2.30 (s, pyrimidine-CH_3), δ 4.41 (s, pyrimidine-CH_2), δ 6.62 (d, J=8.8 Hz, 3,5-Ar-H), δ 6.75 (s, 5-pyrimidine-H), δ 7.71 (d, J=8.8 Hz, 2,6-Ar-H)	13.6	2.3
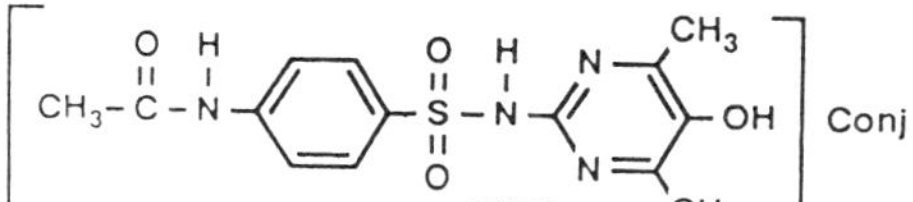 VIII	Analysis of hydrolysis product; EI-MS, m/z 272 (M-SO_2), 230, 214, 139; ^{1}H-NMR (CD_3OD), δ 2.12 (s, O=C-CH_3), δ 2.35 (s, pyrimidine-CH_3), δ 7.62 (d, J=9.2 Hz, 3,5-Ar-H), δ 7.96 (d, J=9.2 Hz, 2,6-Ar-H)	2.5	0.1
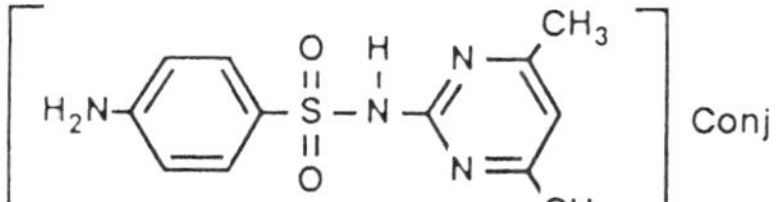 IX	Analysis of hydrolysis product; +FAB-MS, m/z 279 (M+H); ^{1}H-NMR (CD_3OD), δ 2.29 (s, pyrimidine-CH_3), δ 6.63 (d, J=8.8 Hz, 3,5-Ar-H), δ 6.69 (s, 5-pyrimidine-H), δ 7.74 (d, J=8.8 Hz, 2,6-Ar-H)	2.4	0.8
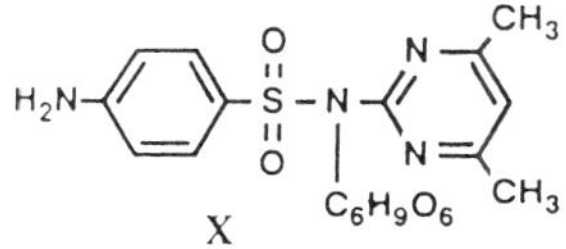 X	-FAB-MS, m/z 453 (M-H); +FAB-MS, m/z 455 (M+H); ^{1}H-NMR (CD_3OD), δ 2.39 (s, pyrimidine-CH_3), δ 3.44-3.72 (complex multiplet, H on C_2-C_5 of uronic acid), δ 5.45 (doublet, J=8.8 Hz, anomeric-H), δ 6.64 (d, J=8.8 Hz, 3,5-Ar-H), δ 6.99 (s, 5-pyrimidine-H), δ 7.78 (d, J=8.8 Hz, 2,6-Ar-H)	4.1	8.9

Continued on next page

Table IV. Continued.

Assigned Structure & Chromatographic Fraction	Evidence for Structure	Route of Dosing Oral	Intravenous
		---(% of ^{14}C in urine)---	
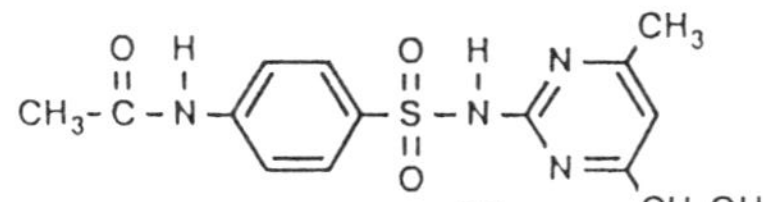 XI	-FAB-MS, m/z 335 (M-H); +FAB-MS, m/z 337 (M+H); ^{1}H-NMR (CD_3OD), δ 2.12 (s, O=C-CH_3), δ 2.24 (s, pyrimidine-CH_3), δ 4.34 (s, pyrimidine-CH_2), δ 6.59 (s, 5-pyrimidine-H), δ 7.59 (d, J=8.4 Hz, 3,5-Ar-H), δ 7.91 (d, J=8.4 Hz, 2,6-Ar-H)	1.4	1.1
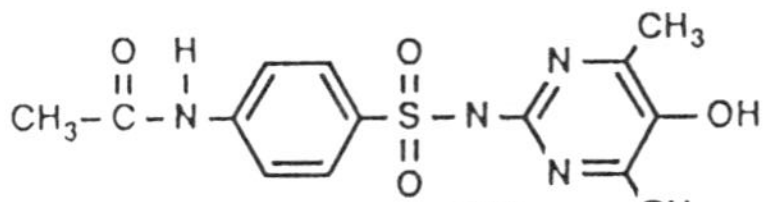 XII	EI-MS, m/z 272 (M-SO_2), 139; -FAB-MS, m/z 335 (M-H); ^{1}H-NMR (CD_3OD), δ 2.13 (s, O=C-CH_3), δ 2.22 (s, pyrimidine-CH_3), δ 7.62 (d, J=8.8 Hz, 3,5-Ar-H), δ 7.92 (d, J=8.8 Hz, 2,6-Ar-H)	0.2	ND
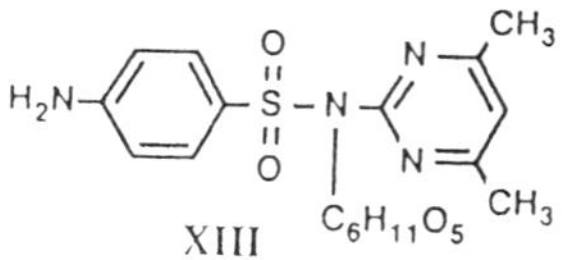 XIII	+FAB-MS, m/z 441 (M+H), 279 (sulfamethazine + H); ^{1}H-NMR (CD_3OD), δ 2.37 (s, pyrimidine-CH_3), δ 3.38-3.91 (complex multiplet, H on C_2-C_6 of hexose), δ 5.43 (d, J=9.2 Hz, anomeric-H), δ 6.65 (d, J=8.8 Hz, 3,5-Ar-H), δ 7.02 (s, 5-pyrimidine-H), δ 7.73 (d, J=8.8 Hz, 2,6-Ar-H)	<.5	<1.0
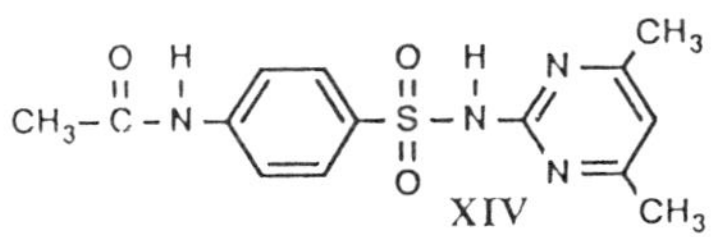 XIV	+FAB-MS, m/z 321 (M+H); ^{1}H-NMR (CD_3OD), δ 2.13 (s, O=C-CH_3), δ 2.29 (s, pyrimidine-CH_3), δ 6.69 (5-pyrimidine-H), δ 7.70 (d, J=8.8 Hz, 3,5-Ar-H), δ 8.00 (d, J=8.8 Hz, 2,6-Ar-H)	22.7	22.8
Unidentified fractions[a]		12.9	6.8

[a]This includes unidentified fractions U-I, U-II, U-III, U-IV and U-V as illustrated in Figure 1.

methazine (Table IV) and the results of systematic ^{1}H-NMR analysis of other substituted aromatic compounds (*19, 20*).

Fraction II was identified as sulfanilamide by comparing its HPLC retention time and its ^{1}H-NMR and mass spectra with those of the reference compound. This minor metabolite was detected in the urine of the intravenously dosed cow but not in the urine of the orally dosed cow. To our knowledge, this is the first report of an animal system cleaving a sulfonamide drug at the N^1-position to yield sulfanilamide. We considered the possibility that ^{14}C-sulfanilamide was a contaminant in the dose given to the cow; however, this seems very unlikely because the retention times of sulfamethazine and sulfanilamide differed by 14 minutes in the HPLC system used to purify ^{14}C-sulfamethazine.

Nielsen (*17*) reported that sulfaguanidine was a minor metabolite in the urine of cows dosed intravenously with sulfamethazine. However, we found no evidence for this compound in the urine from the orally or intravenously dosed cows.

The distributions of ^{14}C-labeled compounds in the tissues (Table III) and urine (Table IV) indicate that the metabolism of ^{14}C-sulfamethazine in the orally and intravenously dosed cows was similar. Four metabolites (fractions II, III, IV and XII) were observed in the urine after only one route of administration; however, each of these metabolites accounted for 1.4% or less of the total ^{14}C-activity in the urine, and it is possible they were present in the urine of the other animal in trace amounts but below the level of detection for the methodology used in these studies. Most of the types of sulfamethazine metabolism observed in these animals has been reported before. However, not previously identified were the metabolites resulting from transformations at two sites on the molecule (e.g. N^4-acetylation and hydroxylation of heterocylic ring; N^4 acetylation and formation of hydroxymethyl group). Also, the ^{1}H-NMR and mass spectral data described here is definitive evidence for hexuronic acid, hexose and sulfate ester conjugates which were previously proposed on the basis of the release of hydroxymethyl and phenolic products by sulfatase or glucuronidase enzymes. Cleavage of ^{14}C-sulfamethazine at the N^1-position to yield sulfanilamide is a new type of metabolism not previously reported. Further studies to determine the mechanism(s) involved in this biotransformation should be considered.

We have not conducted any studies to determine the biological activity(s) of the metabolites reported here. Therefore, we cannot speculate as to whether or not the different types of sulfamethazine metabolism reported here are in any way related to the toxicological responses observed when rodents were dosed with large amounts of sulfamethazine (*1*). Further investigations to evaluate the activity(s), if any, of these metabolites would be of interest.

Mention of trademark or proprietary product does not constitute a guarantee or warranty of the product by the U.S. Department of Agriculture and does not imply its approval to the exclusion of other products that may also be suitable.

Literature Cited.

1. Littlefield, N.A.; Gaylor, D.W.; Blackwell, B.N.; and Allen, R.R. *Food & Chem. Toxicol.* **1989,** *27,* pp. 455-463.
2. Charm, S.E.; Zomer, E.; and Salter R. *J. Food Protection,* **1988,** *12,* pp. 920-924.

3. Clark, E. *Dairy Herd Manage.* **1988,** *25,* p.13.
4. Larocque, L.; Carignan, G.; and Sved, S. *J. Assoc. Off. Anal. Chem.* **1990,** *73,* pp. 365-367.
5. Paulson, G.D.; Giddings, J.M.; Lamoureux, C.H.; Mansager, E.R.; and Struble, C.B. *Drug Metab. Dispos.* **1981,** *9,* pp. 142-146.
6. Paulson, G.D.; Mitchell, A.D.; and Zaylskie, R.G. *J. Assoc Off. Anal. Chem.* **1985**, *68*, pp. 1000-1006.
7. Takahashi, Y.; Said, A.A.; Hashizume, M.; Kido, Y. *J. Vet. Med. Sci.,* **1991,** *53,* pp. 33-36.
8. Vree, T.B. and Hekster, Y.A. *Antibiotics and Chemotherapy* **1985,** *34,* pp. 1-208.
9. Nouws, J.F.M.; Mevius, D.; Vree, T.B.; Baakman, M.; and Degen, M. *Am. J. Vet. Res.* **1988,** *49,*1059-1065.
10. Nouws, J.F.M.; Vree, T.B.; Aerts, R.; and Grondel, J. *Arch. für Lebensmittelhygiene.* **1986,** *37,* pp. 57-74.
11. Nouws, J.F.M.; Vree, T.B.; Aerts, R.; and Grondel, J. In *Agricultural Uses of Antibodies;* ACS Symposium Series 320, **1986** pp. 168-182.
12. Nouws, J.F.M.; Vree, T.B.; Breukink, H.J.; Baakman, M.; Driessens, F.; Smulders, A. *The Vet. Quart.* **1985,** *7,* pp. 177-185.
13. Shetty, S.N. and Asuzu, I.U. *Indian. Vet. J.* **1989,** *66,* pp. 928-932.
14. Kakudo, Y. and Yoshida, Y. *J. Osaka Dental University* **1972,** *6,* pp. 51-87.
15. Atef, M.; El-Sayed, G.A.; Youssef, S.E.A.; El-Gendi, A.Y. and Fadali, Zbl. *Vet. Med. A.* **1981,** *28,* pp. 122-130.
16. Atef, M.; Salem, A.A.; Al-Samarrae; and Zafer, S.A. *Res. Vet. Sci.* **1979,** *27,* pp. 9-14.
17. Nielsen, P. *Biochem. J.* **1973,** *136,* pp. 1039-1045.
18. Bray, H.G.; Lake, H.J.; and Thorpe, W.V. *Biochem. J.* **1951,** *48,* pp. 400-406.
19. Zanger, M. *Org. Magn. Reson.* **1972,** *4,* pp. 1-25.
20. Feil, V.J. In *Xenobiotic Conjugation Chemistry;* Paulson, G.D.; Caldwell, J.; Hutson, D.H.; and Menn, J.J., eds.; ACS Symposium Series 299, **1986,** pp. 177-202.

RECEIVED March 11, 1992

Chapter 14

Ivermectin and Abamectin Metabolism

Differences and Similarities

Bruce A. Halley, Narayana I. Narasimhan, Kalpana Venkataraman, Rae Taub, Marilyn L. G. Erwin, Nicholas W. Andrew, and Peter G. Wislocki

Merck Sharp & Dohme Research Laboratories, Merck & Company, Inc., P.O. Box 2000, Rahway, NJ 07065

The metabolic fates of ivermectin (IVM) and abamectin (avermectin B_1, AVM) were previously studied in our laboratory. The major metabolites of the components of IVM, H_2B_{1a} and H_2B_{1b}, in cattle, sheep and rats were the 24-hydroxymethyl (24-OHMe) compounds, while the major metabolites in swine were the 3"-O-desmethyl (3"-ODMe) compounds. For IVM, there was no identification of 24-OHMe compounds in the swine, no identification of 3"-ODMe compounds in cattle or sheep and evidence for only trace amounts of 3"-ODMe compounds in the rat. For AVM (or its major component B_{1a}), however, both metabolic pathways were evident in all species tested. Hydroxylation of the 24-methyl group was the major metabolic pathway in cattle and goats, while 3"-O-desmethylation was identified as a minor pathway. In rats, 3"-ODMe-B_{1a} was the major metabolite and 24-OHMe-B_{1a} was a minor one. Thus, it appeared there were inter- and intra-species differences in the metabolism of IVM and AVM. Recently, we re-examined the metabolism of IVM and AVM in cattle, swine and rats. We identified 3"-ODMe-H_2B_{1a} in the livers of steers dosed orally with $[^3H]H_2B_{1a}$ and following incubations of $[^3H]H_2B_{1a}$ with steer liver microsomes. We also identified 24-OHMe metabolites in the livers of swine fed $[^3H]$IVM for seven days. In *in vitro* studies of $[^3H]H_2B_{1a}$ and $[^3H]B_{1a}$ with microsomes from livers of untreated steers, swine and rats, similar metabolic profiles were obtained. Also, 3"-ODMe-H_2B_{1a} and 3"-ODMe-B_{1a} were identified as *in vitro* metabolites of IVM and AVM, respectively, in rats. Thus, the rat is a good laboratory animal toxicity model for both cattle and swine. The results of these studies indicate that IVM and AVM are metabolized qualitatively similarly by cattle, swine and rat and that the metabolic profiles for each compound are qualitatively similar among the species.

0097–6156/92/0503–0203$06.00/0

The avermectins are a family of fermentation products possessing a 16-member cyclic lactone, a spiroketal moiety and a disaccharide unit. Avermectin B_1 is a mixture of two components which possess a double bond at position 22,23 and differ by a methylene group in the side-chain substituents at position 25. The mixture containing ≥80% avermectin B_{1a} (B_{1a}) and ≤20% avermectin B_{1b} (B_{1b}) is also designated as abamectin (AVM). Ivermectin (IVM, Figure 1) is produced from AVM by reducing the 22,23-double bond, producing dihydroavermectin B_{1a} (H_2B_{1a}) and dihydroavermectin B_{1b} (H_2B_{1b}). This is the route of introduction of the tritium label into IVM. To label AVM, the hydroxyl group at the 5 position is oxidized to the carbonyl, then reduced with [^{3}H] borohydride (*1*). This [5-^{3}H] label has been shown to be stable in metabolism studies of the B_{1a} component of AVM in rats and goats (*2,3*). IVM is used as an antiparasitic agent in food and companion animals and in man; AVM is also an antiparasitic agent, but is primarily used in agriculture as a acaricide and insecticide (*4*).

Previously, the metabolism of IVM had been studied in cattle, sheep, swine and rats, while the metabolism of AVM had been studied in cattle, goats and rats (*2,3,5-9*). Table I shows a summary of the major (++) and minor (+) *in vivo* (liver) and *in vitro* (liver microsomes) metabolites identified for cattle, rat and swine. Miwa *et al.* (*5*) identified the 24-hydroxymethyl (24-OHMe) compounds as *in vitro* metabolites of IVM and AVM in incubations with rat and steer liver microsomes. Chiu *et al.* (*7,8*) confirmed 24-OHMe-H_2B_{1a} was the major *in vivo* metabolite of IVM in the livers of cattle and rats, while 24-OHMe-H_2B_{1a}-monosaccharide (24-OHMe-H_2B_{1a}-MS) was identified as a minor metabolite in both species. In the swine, however, little radioactivity eluted in the reversed-phase high performance liquid chromatography (RP-HPLC) regions associated with 24-OHMe-H_2B_{1a} and 24-OHMe-H_2B_{1a}-MS when the extract of a composite of livers from swine, sacrificed 14- and 21-days post a single subcutaneous 400 µg/kg dose of [^{3}H]IVM, was chromatographed (*8*). Instead, the major metabolite was 3"-ODMe-H_2B_{1a}. This was confirmed *in vitro* using swine liver microsomes (*6*). This metabolite had not been identified in the steer or rat, though there were "drug-like" metabolites which coeluted from RP-HPLC with H_2B_{1b} (where 3"-ODMe-H_2B_{1a} would be expected).

For AVM, 24-OHMe-B_{1a} was the major *in vivo* metabolite of AVM in cattle liver at 14 days post a 300 µg/kg subcutaneous dose (Chiu, S.H.L, Merck & Co., Inc., personal communication, 1986). Maynard *et al.* (*3*) identified 24-OHMe-B_{1a} (3 - 12% of the total residue) in livers of rats dosed with [5-^{3}H]B_{1a}, but it was not the major metabolite. The major metabolite was 3"-ODMe-B_{1a} (20 - 40%). These results were also reflected in *in vitro* incubations where, with 2% overall metabolism, 3"-ODMe-B_{1a} and 24-OHMe-B_{1a} comprised 1.5% and 0.5%, respectively.

Thus, there appeared to be differences between the metabolism of IVM and AVM in cattle, rat and swine. Cattle and rat appeared to metabolize IVM similarly, but differently than swine. Previous work indicated the presence of radioactivity in the HPLC eluates in the 3"-ODMe-H_2B_{1a} region for cattle and "polars" (24-OHMe-B_{1a} region) for swine, but these metabolites were not further identified. Cattle appeared to metabolize IVM and AVM similarly, but cattle and rat appeared to metabolize AVM differently. No data were available on the metabolism of AVM by swine.

We reinvestigated the metabolism of IVM while comparing dosage routes other than subcutaneous. The first study involved an intraruminal route with cattle; the second involved an oral route with swine. We also compared the *in vitro* metabolism of IVM and AVM by liver microsomes from control steers, rats and swine.

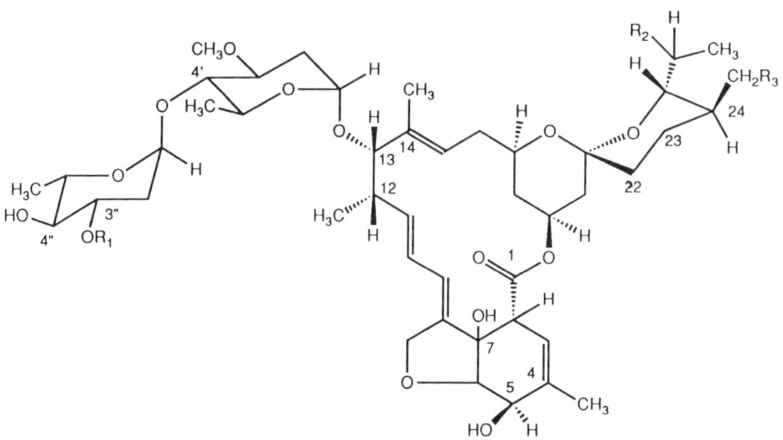

	$\underline{R_1}$	$\underline{R_2}$	$\underline{R_3}$
H_2B_{1a}	CH_3	CH_2CH_3	H
H_2B_{1b}	CH_3	CH_3	H
24-Hydroxymethyl H_2B_{1a}	CH_3	CH_2CH_3	OH
3"-O-Desmethyl H_2B_{1a}	H	CH_2CH_3	H

Figure 1. The structure of ivermectin (>80% H_2B_{1a}, <20% H_2B_{1b}) and the major metabolites of H_2B_{1a}.

Table I. Major and Minor Metabolites of IVM and AVM Identified in Previous Studies

	IVM			AVM	
	24-OHMe -MS	24-OHMe	3"-ODMe	24-OHMe	3"-ODMe
CATTLE	+	++		++	
RAT	+	++		+	++
SWINE			++		

++ = major metabolite
\+ = minor metabolite

Experimental

Tissue extractions, microsomal preparations and incubations, reversed-phase (RP-HPLC) and normal-phase high performance liquid chromatography (NP-HPLC) were conducted similarly to previously described methods (*5, 6, 10*). HPLC-detection was by absorbance at 245 nm and by scintillation counting of 1 minute eluate fractions. Nuclear magnetic resonance spectroscopy (nmr) and mass spectroscopy (ms) conditions were similar to those described previously (*6*). For the study where IVM was dosed intraruminally to cattle, three steers, 200 - 300 kg body weight and <1 year old, were each given a single intraruminal dose of 200 μg/kg of [22,23-^{3}H]H_2B_{1a} (93μCi/mg in propylene glycol). Three steers served as untreated controls. After 7 days, the animals were sacrificed and the livers were taken. For the studies where IVM was fed to swine, male (barrows) and female (gilt) pigs, ~30 kg body weight and <4 months old, were dosed with [22,23-^{3}H]H_2B_{1a} and [22,23-^{3}H]H_2B_{1b} (93%: 7%), each at 300 μCi/mg, at 100 μg of IVM per kg body weight per day in their feed for 7 days. Feeding was twice a day and the pigs were sacrificed 4 hours post last medicated feeding. For the incubations of IVM and AVM with microsomes from livers from non-induced rat, steer and swine, the incubation system consisted of 1.7 mg (1 μCi) of [22,23-^{3}H]H_2B_{1a} or [5-^{3}H]B_{1a} in 0.2 mL of methanol added to 4 mg of microsomal protein and an NADPH generating system in 1.2 mL of pH 7.4 phosphate buffer. Using p-nitroanisole as a substrate, the cytochrome P-450-mediated O-demethylase activity was about 1 nmol product formed per min per mg protein for all microsomal preparations. Two control incubations were included for each substrate; one contained no NADPH generating system (buffer added in its place) and the other contained heat-inactivated microsomes (10 min at 80 °C). The samples were incubated at 37 °C. More buffer, microsomes and generating system were added at 2, 4, 6 and 8 hours. The incubation mixtures were passed through C-18 solid phase extraction cartridges. After rinsing the cartridges with water, the radioactivity was eluted with methanol. The samples were concentrated and chromatographed using RP-HPLC.

Results and Discussion

IVM Dosed Intraruminally to Cattle. A representative RP-HPLC of the extract from one of the 7-day post-dose livers from a treated steer is shown in Figure 2 (top). The unchanged H_2B_{1a} (fractions 40 - 45) comprised ~58% of the eluted drug residue, while the polar metabolites (fractions 3 - 9) comprised ~32%. About 7% of the drug residue eluted before the H_2B_{1a}, where unlabeled H_2B_{1b}, added as "cold carrier", eluted. Since labeled H_2B_{1b} was not used in this study, this "drug-like" metabolite was suspected to be 3"-ODMe-H_2B_{1a} which under these HPLC conditions also elutes at the retention time of H_2B_{1b}. This identification was strengthened by NP-HPLC (Figure 2, bottom) where ~8% of the drug-residue eluted where 3"-ODMe-H_2B_{1a} would be expected (*10*). Similar results were observed when [22,23-^{3}H]H_2B_{1a} was incubated with liver microsomes prepared from the control steers. Again, by both RP- and NP-HPLC, radioactive metabolites eluted where 24-OHMe-H_2B_{1a} and 3"-ODMe-H_2B_{1a} would be expected (data not shown).

The feces were also collected from the dosed steers. Acetonitrile extracts from day 2 and day 6 post dose feces were chromatographed on both RP- and NP-HPLC systems. At 2 days post dose, ~94% of the drug-residue was H_2B_{1a}. By day 6, the total residue level in the feces had dropped ~25-fold relative to day 2 (~80 ppb and ~2 ppm respectively) and the percent of H_2B_{1a} decreased to ~50%. Based on RP- and NP-HPLC elution times, 24-OHMe-H_2B_{1a} accounted for ~36% and 3"-ODMe-H_2B_{1a} accounted for ~12% of the residue in the extract of day 6 feces.

Thus, this study indicated that steers produced 3"-ODMe-H_2B_{1a} in addition to the major metabolite, 24-OHMe-H_2B_{1a}, and that 3"-ODMe-H_2B_{1a} could also be detected in *in vitro* incubations using microsomes from control steer livers.

IVM Fed to Swine. The RP-HPLC of an extract of liver from swine, Figure 3, shows the UV absorbance (top) associated with the added cold-carrier IVM (36 - 58 min). About 34% of the radioactivity (bottom) coeluted with H_2B_{1a} (48 - 58 min), while ~40% coeluted with H_2B_{1b} (36 - 41 min). Polar metabolites (4 - 8 min) comprised ~15% of the total, while at least 2 "drug-like" metabolites (28 - 34 min) comprised another ~10%. The radioactivity in the drug region is not proportional to the percentages of [22,23-^{3}H]H_2B_{1a} and [22,23-^{3}H]H_2B_{1b} in the dosing mixture, ~92:8. Use of just RP-HPLC data would lead to a wrong conclusion about the percent of H_2B_{1a} and H_2B_{1b} in the liver from the on-drug animal but, based on studies of swine dosed subcutaneously with IVM (*9*), NP-HPLC was used to separate the H_2B_{1b} from the coeluting 3"-ODMe-H_2B_{1a}. Figure 4 shows the NP-HPLC of the same extract. By NP-HPLC, the drug region (H_2B_{1a} and H_2B_{1b}) comprised ~40% of the total drug residue, the 3"-ODMe and drug-like region comprised ~40% and the polars, where 24-OHMe-metabolites elute, comprised ~15%. Each region was collected and rechromatographed on RP-HPLC, Figure 5. The drug region is ~ 16% H_2B_{1b} and ~ 77% H_2B_{1a} (or ~7% and ~32% overall), while the 3"-ODMe and drug-like region contained at least 4 compounds. The major metabolite in the 3"-ODMe and drug-like region was isolated and characterized as 3"-ODMe-H_2B_{1a} by cochromatography with an authentic sample which was generated *in vitro* and identified by ms and nmr. The identity of the polar

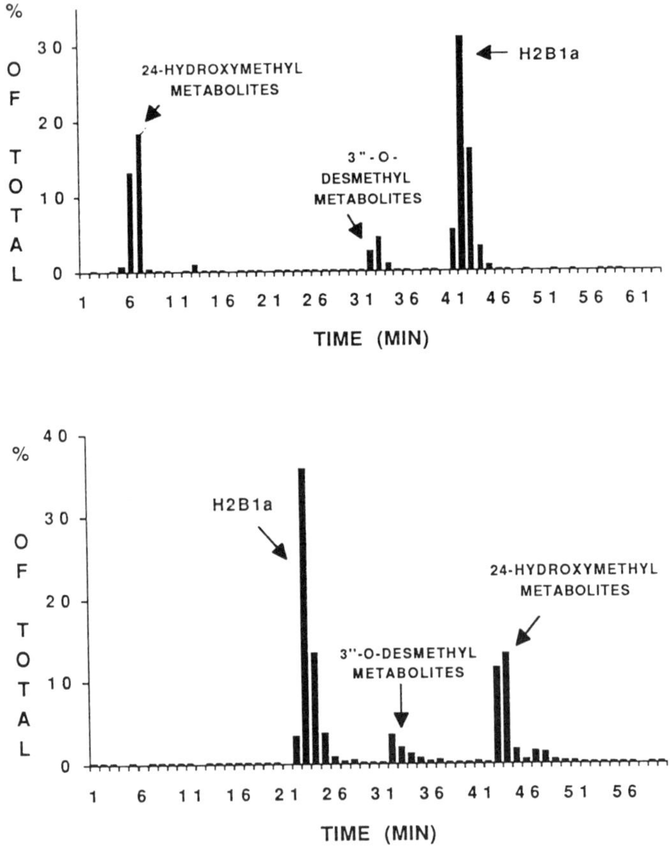

Figure 2. (top) RP-HPLC histogram of the eluted radioactivity from an extract from the liver of a steer dosed intraruminally with [22,23-^{3}H]H_2B_{1a}, 7 days post dose. (bottom) NP-HPLC histogram of the eluted radioactivity from the same extract.

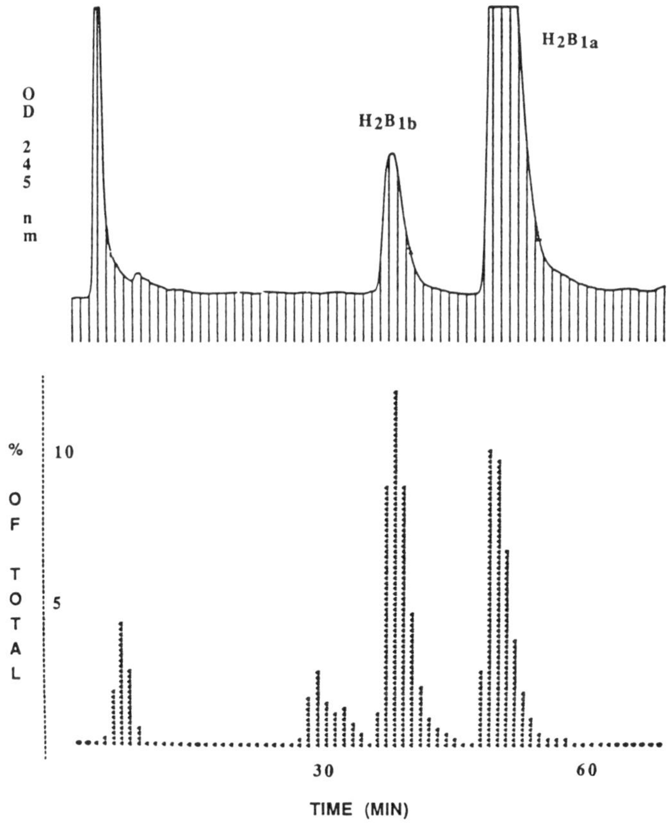

Figure 3. RP-HPLC histogram of the eluted radioactivity from an extract from the liver of a swine fed [22,23-^{3}H]IVM in medicated feed for 7 days. (top, UV; bottom, radioactivity)

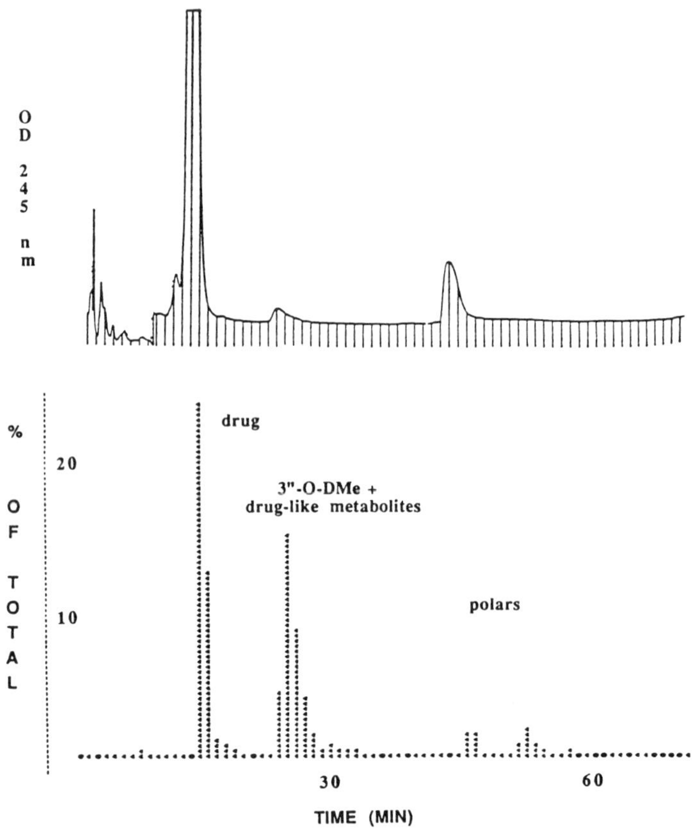

Figure 4. NP-HPLC histogram of the eluted radioactivity from an extract from the liver of a swine fed [22,23-^{3}H]IVM in medicated feed for 7 days. (top, UV; bottom, radioactivity)

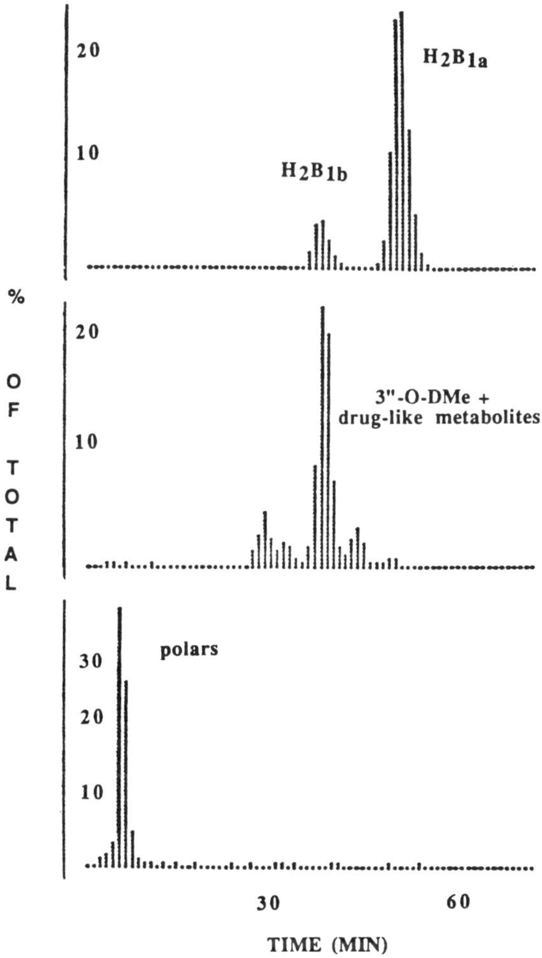

Figure 5. RP-HPLC histogram of NP-HPLC fractions from an extract from the liver of a swine fed [22,23-^{3}H]IVM in medicated feed for 7 days, see Figure 4. (top) Drug region, (middle) 3"-O-DMe + drug-like region and (bottom) polars region from the NP-HPLC.

metabolites was further examined using a second RP-HPLC system which utilized 40% H_2O (7). Using these conditions, ~7% of the total radioactivity in the extract from the on-drug liver eluted in the 24-OHMe-H_2B_{1a} region. Further confirmation of 24-OHMe-H_2B_{1a} as a metabolite of IVM in swine was obtained by chromatographing an extract of swine feces collected 1 day post last dose. Approximately 45% of the drug-residue in this feces sample was comprised of polar metabolites, and ~37% of the drug-residue eluted in the 24-OHMe-H_2B_{1a} region using the 40% H_2O RP-HPLC system.

Thus, using two RP-HPLC systems and one NP-HPLC system, this study established 24-OHMe-H_2B_{1a} as a metabolite of IVM in swine liver and feces.

Incubations of IVM and AVM with Microsomes from Livers from Non-induced Rat, Steer and Swine. Since the previous studies (above) used only chromatography or cochromatography to characterize the metabolites, a study was conducted to isolate enough material for identification by spectroscopic methods. IVM and AVM were metabolized, in a side-by-side *in vitro* comparison, using microsomes prepared from the livers of untreated rats, steers and swine. There was no metabolism in any of the control incubations, i.e., those using heat-inactivated microsomes or no NADPH generating system. The observed chromatograms were corrected for trace impurities in the starting radiolabeled materials by subtracting the eluted radioactivities in each corresponding fraction of the control samples excluding the fractions containing IVM or AVM.

Figures 6 and 7, respectively, compare the radioactivity eluting from the RP-HPLC of the extracts from the IVM and AVM incubations. The peaks from [22,23-^{3}H]H_2B_{1a} and [5-^{3}H]B_{1a} are off-scale in each trace to more clearly show the peaks from the metabolites. In each incubation, the percent of metabolism was low, making quantitative comparisons difficult. However, qualitative comparisons can be made, and there was sufficent conversion for isolation and identification of some of the metabolites. In Figure 6, H_2B_{1a} eluted at ~30 min. Two metabolite regions can be seen; the 24-OHMe region just after the column void volume and the 3"-ODMe region eluting just prior to the H_2B_{1a}. As can be seen, the incubations with rat, steer and swine microsomes all produced both metabolites. Similarly, the same microsomes all produced 24-OHMe- and 3"-ODMe-metabolites from AVM, Figure 7.

The material in the 3"-ODMe-region from the incubation of [22,23-^{3}H]H_2B_{1a} with rat microsomes was isolated for spectroscopic identification. It had an absorbance maximum of 246 ± 2 nm, consistent with the IVM chromophore (*4*), a single nmr peak with an integrated intensity of 3 at 3.45δ, unlike IVM which has two singlets in this region, and a molecular weight of 860 based on m/z ratios of 867 and 883, respectively, using FAB-ms and spiking with lithium and sodium salts. The nmr and ms data are consistent with demethylation, therefore, the rat microsomes did produce 3"-ODMe-H_2B_{1a}, just as the steer and swine. Likewise, the material in the 3"-ODMe-region from the incubation of [5-^{3}H]B_{1a} with swine microsomes was isolated for spectroscopic identification. It had an absorbance maximum of 244 ± 2 nm, consistent with the AVM chromophore (*4*), and a single nmr peak with an integrated intensity of 3 at 3.44δ. It cochromatographed (RP-HPLC) with the major metabolite from the incubation of [5-^{3}H]B_{1a} with the rat liver microsomes, which was previously identified as 3"-ODMe-B_{1a} (*3*). Thus, the swine microsomes produced 3"-ODMe-B_{1a} as the major metabolite and material in the 24-OHMe region as minor metabolites.

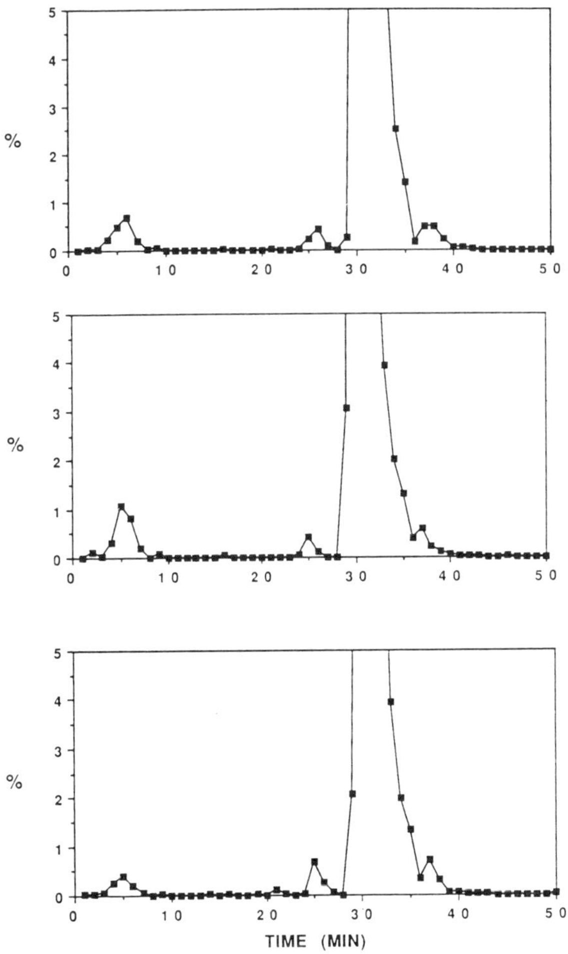

Figure 6. *In vitro* IVM Metabolites: RP-HPLC histograms of the eluted radioactivity from extracts from the incubations of [22,23-^{3}H]H_2B_{1a} with rat (top), steer (middle) and swine (bottom) liver microsomes.

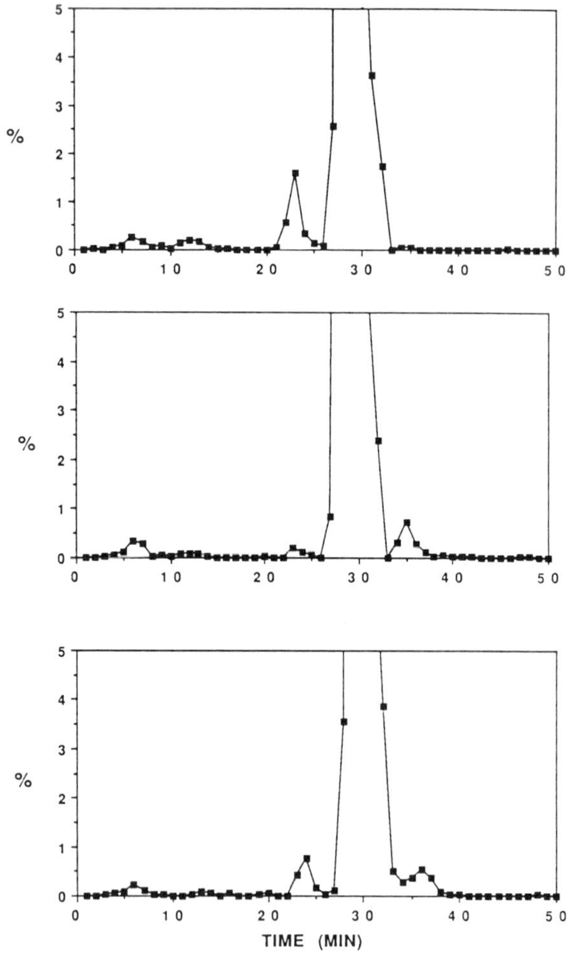

Figure 7. *In vitro* AVM Metabolites: RP-HPLC histograms of the eluted radioactivity from extracts from the incubations of [5-^{3}H]B_{1a} with rat (top), steer (middle) and swine (bottom) liver microsomes.

Summary

Table II summarizes the results of the previous and recent metabolism studies with IVM and AVM. The identifications of the major (**) and minor (*) *in vivo* (liver) and *in vitro* (liver microsome) metabolites determined for cattle, rat and swine in these recent studies are included with the data from Table I. The metabolism of IVM and AVM is qualitatively similar for all three species and *in vivo* and *in vitro* metabolism is similar for each species. There are quantitative differences between species (compare the metabolism of IVM by rat and cattle with that by swine) and between compounds for a given species (compare the metabolism of IVM and AVM by rat), but all three species produce either 24-OHMe- or 3"-ODMe-metabolites as the major metabolite and the other compound as a minor metabolite. The rat is therefore an appropriate laboratory animal toxicity model for both cattle and swine for IVM and AVM.

Table II. Major and Minor Metabolites of IVM and AVM Including Recent *in vivo* and *in vitro* Results

	IVM			AVM	
	24-OHMe -MS	24-OHMe	3"-ODMe	24-OHMe	3"-ODMe
CATTLE	+	++	*	++	*
RAT	+	++	*	+	++
SWINE		*	++	*	**

++ = previously identified major metabolite
\+ = previously identified minor metabolite
** = major metabolite, identified in current studies
* = minor metabolite, identified in current studies

Acknowledgments. The preparation of the radiolabeled materials by the Labeled Compound Synthesis Group of MSDRL, the dosing and handling of the cattle and swine and the tissue preparation by Animal Metabolism/Branchburg Farm and the generation and interpretation of the nmr and ms data by Drs. Byron Arison and Lawrence Colwell are all greatly appreciated.

Literature Cited

1. Chabala, J.C.; Rosegay, A.; Walsh, M.A.R. *J. Agric. Food Chem.* **1981**, *29*, 881-884.
2. Maynard, M.S.; Wislocki, P.G.; Ku, C.C. *J. Agric. Food Chem.* **1989**, *37*, 1491-1497.

3. Maynard, M.S.; Halley, B.A.; Green-Erwin, M.; Alvaro, R.; Gruber, V.F.;Hwang, S.C.; Bennett, B.W. Wislocki, P.G. *J. Agric. Food Chem.* **1990**, *38*, 864-870.
4. *Ivermectin and Abamectin*; Campbell, W. C., Ed.; Springer-Verlag: New York, NY, 1989.
5. Miwa, G.T.; Walsh, J.S.; VandenHeuvel, W.J.A.; Arison, B.; Sestokas, E.; Buhs, R.; Rosegay, A.; Avermitilis, S.; Lu, A.Y.H.; Walsh, M.A.R.; Walker, R.W.; Taub, R.; Jacob, T.A. *Drug Metab. Dispos.* **1982**, *10*, 268-274.
6. Chiu, S.H.L.; Sestokas, E.; Taub, R.; Smith, J.L.; Arison, B.; Lu, A.Y.H. *Drug Metab. Dispos.* **1984**, *12,* 464-469.
7. Chiu, S.H.L.; Sestokas, E.; Taub, R.; Buhs, R.P.; Green, M.; Sestokas, R.; VandenHeuvel, W.J.A.; Arison, B.H.; Jacob, T.A. *Drug Metab. Dispos.* **1986**, *14,* 590-600.
8. Chiu, S.H.L.; Taub, R.; Sestokas, E.; Lu, A.Y.H.; Jacob, T.A. *Drug Metab. Dispos.* **1987**, *18,* 289-302.
9. Chiu, S.H.L.; Sestokas, E.; Taub, R.; Green, M.; Baylis, F.P.; Jacob, T.A.; Lu, A.Y.H. *J. Agric. Food Chem.* **1990**, *38,* 2079-2085.
10. Chiu, S.H.L.; Sestokas, E.; Taub, R. *J. Chromatogr.* **1988**, *433,* 217-224.

RECEIVED February 28, 1992

Chapter 15

Luprostiol in Dairy Cows

Residue Depletion and Metabolism

C. A. Weerasinghe, E. Warrakah, S. Shuman, and R. Wang

SmithKline Beecham Animal Health, 1600 Paoli Pike, West Chester, PA 19380

Luprostiol is a prostaglandin derivative of the type PGF_{2a} which was developed as a potential agent for estrus synchronization in dairy cows. The depletion of ^{35}S-Luprostiol residues in milk, liver, kidney, fat and muscle was investigated 12 hour, 1 day and 3 days after intramuscular administration of the drug. At a dosage of 15 mg/head, injection site had a residue level of 250 ppb which was the highest tissue residue level detected at 1-day withdrawal. However, the residues were quickly depleted to 11 ppb within 3 days. The residue levels in milk were less than 2 ppb in the first 12 hour sampling period. The metabolic profile in urine was studied by HPLC/MS after a 30 mg/head dose. Several metabolites were detected. The major pathway of degradation was found to be the β-oxidation of the carboxylic acid side chain. The tissue residue levels at different withdrawal periods, and the metabolic profiles in urine, liver and kidney are presented.

Prostaglandins are known to occur ubiquitously in mammals. Among these, prostaglandin F_{2a} type has been extensively studied in association with its use in causing regression of the corpus luteum (1-3). Synchronization of estrus and ovulation in cows have been achieved by discriminate administration of PGF_{2a}. The mechanisms of bovine luteolysis have been described in detail (4). The estrus cycle synchronization has an obvious advantage when used in conjunction with artificial insemination in cows.

Absorption, excretion and tissue residues of synthetic prostaglandin analogs of the PGF_{2a} type: cloprostenol and fenprostalene have been studied in rat and marmoset (5), feedlot heifers (6) and dairy cows (7). In all studies, it was evident that the parent drug was cleared rapidly with no persistent residues of the compound in tissues. Metabolic pathways of PGF_{2a} have been discussed (8). Several studies have been reported where the metabolic pathways of PGF_{2a} analogs, chlorprostenol (9), alfaprostol (10) and arbaprostol (11) have been investigated in dairy cows and rats. Gas chromatographic and combined gas

0097-6156/92/0503-0217$06.00/0

chromatographic and mass spectrometric (GC/MS) methods have been utilized in the identification of metabolites. The application of thermospray HPLC/MS as an alternative method for prostaglandin analysis has been reported (12).

Experimental Methods

Acclimation, Dose Administration and Sacrifice. The study was conducted in several phases involving ten Holstein dairy cows (1000-1500 lb) in their mid-lactation. After two weeks of acclimation in floor pens, animals selected for dosing in each phase of the study were placed in metabolism cages for one week prior to dosing. The selection was based on feed consumption, milk production and clinical condition. The selected animals were prepared by marking the injection sites, usually one to two days prior to the day of dose. Several animals (#59, #704, #821 and #979) were catheterized prior to the administration of the test substance in order to collect urine and feces separately for metabolism and balance excretion studies. The catheter was usually inserted one to two days prior to the injection. The dosage of 15 mg of ^{35}S-Luprostiol in propylene glycol (injection #1) was administered intramuscularly on the left flank injection site (already marked). The second dosage (injection #2) of 15 or 30 mg was given 11 days later. This dose was administered on the right flank. Three withdrawal times were selected for sacrifice and tissue collection. Animals sacrificed after 12, 24, and 72 hours of the second dosage were designated as 0-day, 1-day, and 3-day, respectively. The animals selected for 0-day withdrawal (cows #704, #821, #51 and #59) received approximately 30 mg of the test substance at injection #2. The cows had access to good quality alfalfa hay, corn silage, and fresh water *ad libitum*. The animals were sacrificed using a captive bolt pistol. In order to restrain the animal, an I.M. injection of 3% succinylcholine chloride (50 mL) was given.

Collection of Tissues and Body Fluids. After exsanguination, the tissues were removed in the following order: subcutaneous and abdominal (omental) fat, liver, kidney, flank muscle injection site #1 and finally, injection site #2. The flank muscle tissue was taken from the left side, a few inches (6-10 inches) away from the injection site #1. For each muscle tissue, a 10 cm (diameter) x 6 cm (depth) disc of approximately 500 g, (excluding skin and subcutaneous tissue) was taken. All tissues were removed within two hours of the sacrifice.

Milk samples were collected at 12 hour intervals post-administration (after injection #1) through 168 hours. Pre-dose (one to two) samples were also collected as controls, one being the sample collected immediately prior to dosing. Milk samples were also collected after injection #2 at 12 hour intervals until sacrifice.

Urine and feces samples were collected separately from the

catheterized animals (#704, #821, #51 and #59) for metabolic studies and also to determine the excretion of ^{35}S residues (#704 and #821) in each passage. For metabolic studies, urine collected after 30 mg dose (injection #2) was used. Samples were collected at 12 hour intervals post-administration through 168 hours. Pre-dose (one to two) samples were also collected and used as controls. The urine collected was kept frozen and away from light in order to prevent any possible photolytic degradation of Luprostiol or its metabolites.

Analytical Methods.

Radioassay. All samples were analyzed by liquid scintillation counting. The milk and urine samples were analyzed directly after dissolving in Instagel® which is the scintillation cocktail used. The tissues were digested and decolorized with potassium hydroxide and hydrogen peroxide, respectively before analysis. All samples were mixed or shaken thoroughly and dark adapted in appropriate scintillation cocktails (Atomlight® - two hours, Instagel® - six hours) before radioassay.

Method Validation. The assay methods for tissues and milk were validated by analyzing control tissue and milk samples fortified with ^{35}S-Luprostiol at levels encompassing the appropriate residue levels detected in dosed animals. The detection limit of the assay (which was dictated by the specific activity of the drug) gradually changed as a function of ^{35}S decay during the study.

Metabolite Extraction.

Urine. Urine samples collected from cows #821 and #59 in the 0-12 hour period after the administration of a 30 mg dose (injection #2) were utilized in the isolation and identification of metabolites. The isolation procedure involved ethyl acetate extraction of urine followed by solid phase extraction of the ethyl acetate extract on C-18 Bond Elut® cartridges. The extract was derivatized into methyl esters with diazomethane and subjected to silica and alumina chromatography for further cleanup prior to analysis. The detailed procedure is described in flow diagram 1.

Liver and Kidney. Liver and kidney tissues collected after the sacrifice of cow #59 (0-day withdrawal) were utilized for isolation of metabolites. The isolation procedure involved acetonitrile extraction followed by solid-phase (C-18 Bond Elut® cartridge) extraction of the extracts. Sample cleanup after solid-phase extraction was similar to that of urine. The procedure is described in flow diagram 2.

HPLC Analysis. Urine extracts from cows #821 and #59 were analyzed to compare the metabolic profiles. The extracts of liver and kidney were also analyzed to compare their metabolic profiles with those of urine. The analyses were carried out on a reverse

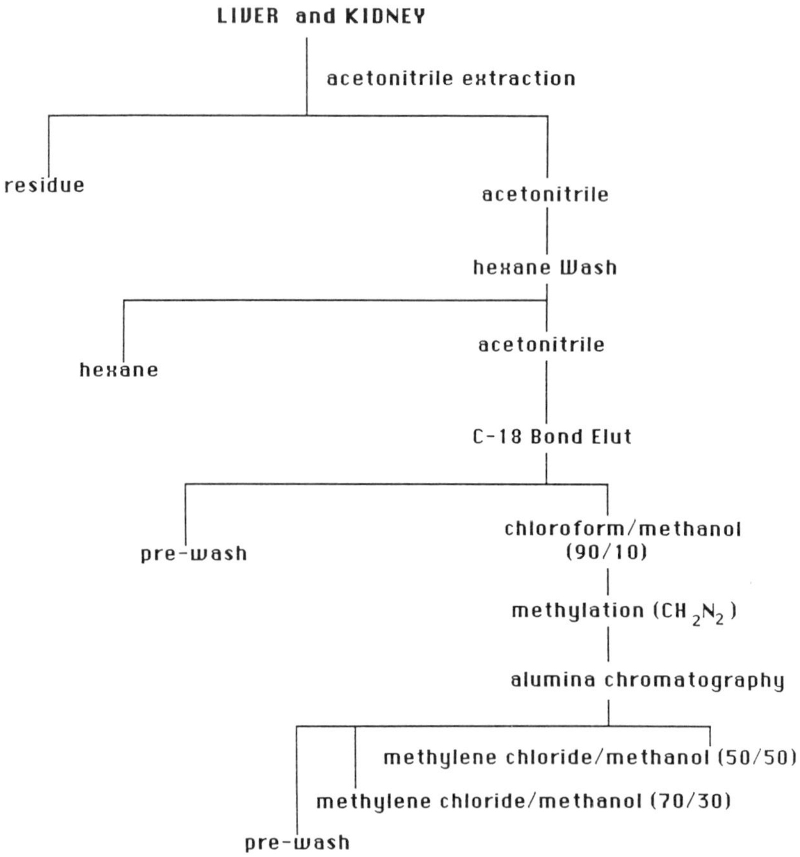

FLOW DIAGRAM 1. CLEANUP PROCEDURE FOR LIVER AND KIDNEY TISSUES

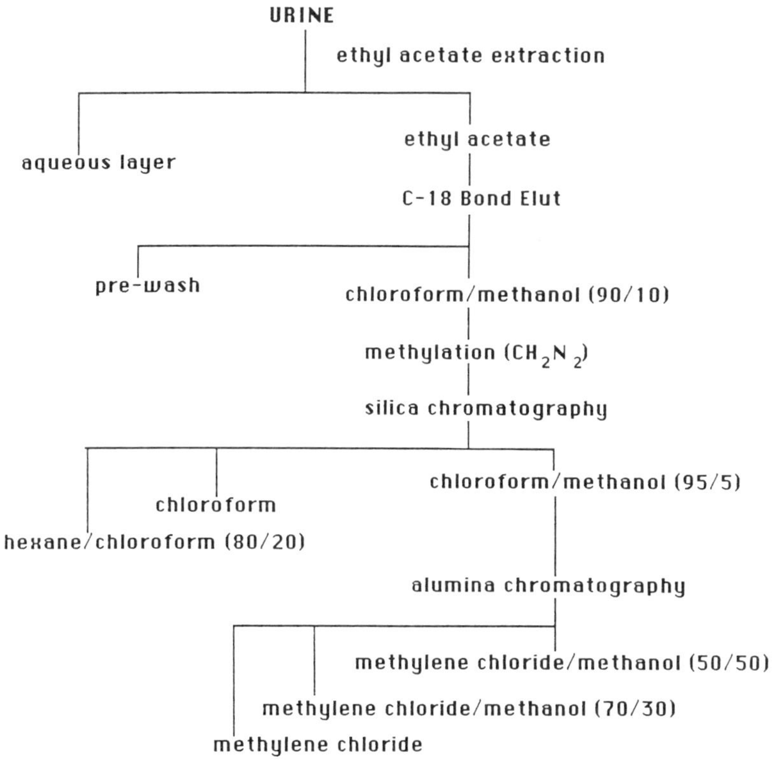

FLOW DIAGRAM 2. CLEANUP PROCEDURE FOR URINE

phase HPLC system with UV and radioactivity detectors connected in series. The HPLC system used is given as footnotes in the figures. After equilibrating the HPLC column with the appropriate mobile phase, a ^{35}S-Luprostiol standard solution was injected to determine the column performance. All sample extracts of urine, liver and kidney were reconstituted in chloroform for final analysis.

Thermospray HPLC/MS. Urine extracts were analyzed by thermospray/HPLC/MS to identify the metabolites present. A Finnigan 4500 series mass spectrometer interfaced with a Vestec Thermospray unit was used to perform the analysis. The aerosol thermocouple of the interface was replaced with a repeller which was connected to an external power supply. The samples were analyzed under thermospray/filament chemical ionization with pulsed +/- ion detection. A mobile phase consisting of acetonitrile/0.1 M ammonium acetate with a linear gradient of 9% to 65% over 30 minutes was used in the anlaysis.

Results and Discussion

Residues in Milk and Tissues. The residues of ^{35}S-Luprostiol and/or its equivalent in milk during the first 48 hours after administration in all 10 animals studied are given in Table I. The residues depleted to 0.04 ppb (mean) 48 hours after administration. After the first dose administration (injection #1), the mean concentration of ^{35}S residues in the 0-12 hour period was 1.81 ± 0.61 ppb (Table I). The total ^{35}S residues excreted in milk up to 48 hours after administration was less than 1% of the total dose administrated. The concentration of ^{35}S-Luprostiol and/or its equivalent in milk in the first 12 hours after the administration of injection #2 are given in Table II. The mean value (4.22 ppb) for 0-day animals was computed separately from the others, since these animals received twice (30 mg) the normal dosage. The ppb values presented were corrected for the accuracy (or efficiency) of the assay method. For milk, the accuracy of the assay method was 95.23 ± 7.06%.

The tissue residue levels are summarized according to the withdrawal-times in Table III. The mean ± standard deviations of the residue levels for all the animals are presented. For liver, kidney and muscle, the assay method for levels above 4 ppb were validated. For fat, the lowest level at which the method was validated is 0.50 ppb (at an accuracy of 48:13 ± 10.54%). The highest residue level was found in the injection site (average 3104.30 ppb) followed by the kidney (average 31.06 ppb) in 0-day animals after a 15 mg and then a 30 mg doses. However, the residue levels were found to be much lower in the injection site (average 10.68 ppb) and kidney (average 0.85 ppb) in the 3-day animals which received two 15 mg doses. There was rapid depletion of ^{35}S residues in all three tissues up to three days after administration.

Table I. RESIDUES (PPB AND % OF TOTAL DOSE) OF LUPROSTIOL AND/OR ITS EQUIVALENT IN MILK (0–48 HOURS) AFTER INJECTION #1 (DOSE #1)

Sampling Time	Animal ID										Concentration (PPB)	% of Total Dose
	#704	#821	#51	#59	#47	#50	#57	#978	#48	#60	Mean±S.D.	Mean+S.D.
12 hours	2.68	2.99	1.86	1.89	2.02	1.31	1.42	1.35	1.51	1.11	1.81±0.61	0.13±0.04
24 hours	0.75	1.13	0.24	0.49	0.69	0.26	0.47	0.50	0.49	0.07	0.51±0.30	0.04±0.02
36 hours	0.21	0.27	0.05	0.09	0.17	0.04	0.12	0.42	0.13	0.04	0.15±0.12	0.01±0.005
48 hours	0.06	0.07	0.01	0.03	0.04	0.02	0.05	0.04	0.02	0.02	0.04±0.02	Trace[a]

[a]Less than 0.01% of the total dose administered.

Table II. RESIDUES (PPB) OF LUPROSTIOL AND/OR ITS EQUIVALENT IN MILK (0–24 HOURS) AFTER INJECTION #2 (DOSE #2)

Sampling[a] Time	#704[b]	#821[b]	#51[b]	#59[b]	#47	#50	#57	#978	#48	#60
12 hours	4.84	5.20	3.22	3.62	1.60	1.65	1.76	1.20	1.21	1.23
	Mean±S.D. = 4.22±0.95				Mean±S.D. = 1.44±0.26					
24 hours					0.30	0.41	0.60	0.48	0.62	0.24
					Mean±S.D. = 0.44±0.16					

[a]Sampling time is the time lapsed after dose administration.

[b]These animals received 30 mg dose which is twice the normal dose.

Excretion of ^{35}S Residues. The excretion of the test substance or its equivalent in the feces and urine was studied in the catheterized animals (cow #704 and #821). It was evident that a majority of the residues (60-90%) was excreted in the first 24 hours post-administration, and within 48 hours, 78-96% of the residues was excreted.

HPLC with radioactivity monitoring indicated eight radioactive peaks in the urine collected in the first sampling period (0-12 hour) after dose administration. For comparison, HPLC profiles of the urine extracts from cow #821 and cow #59 are given in Figure 1. There is only partial resolution between peaks 4, 5 and 6. The overall profile has not changed between the two animals except for some variations in the total percentages of the parent compound (peak #8) and the metabolites. There are at most five major (> 10% of the total radioactivity) components in the urine extract.

Identification of Metabolites of Luprostiol in Urine by Thermospray/HPLC/MS.

The extracts of 0-12 hour urine from cow #821 after the administration of 30 mg dose (Injection #2) was analyzed by thermospray/HPLC/MS. The spectra were acquired under positive and negative ion modes simultaneously under thermospray conditions. However, no useful information was obtained from the spectra acquired under positive ion mode of acquisition. Negative ion spectra showed several characteristic ions with their masses corresponding to $(M+CH_3COO)^-$, $(M+CH_3COO-H_2O)^-$ and $(M+CH_3COO-2H_2O)^-$ as illustrated by the spectrum of standard Luprostiol methyl ester (Figure 2). The molecular ions (acetate adducts) detected during the analysis of urine extracts with their relative retention times (RRTs), which are the retention times of any given peak in the total ion current relative to the retention time of the Luprostiol methyl ester peak, are given in Table IV. The retention time of the latter was confirmed by the analysis of standard Luprostiol methyl ester under same conditions.

For peak assignments, the on-line UV trace obtained during the thermospray/HPLC/MS analysis was not useful due to high background. A radioactivity detector could not be connected on-line due to possible back pressure that might have damaged the flow-cell. Nevertheless, the RRTs of the radioactivity peaks obtained from an off-line HPLC analysis of the same extract(s) were compared with thermospray/HPLC/MS RRTs for structural assignments to the HPLC peaks (Table IV). The following molecular ions were assigned to the radioactivity peaks in the HPLC profile (Figure 3).

Peak #1 (RRT 0.64): This is an early eluting peak and none of the thermospray/HPLC/MS spectra corresponded to its RRT. No further work was pursued since this peak represented 2.45 - 5.59% of all the metabolites.

Table III. TISSUE RESIDUES OF LUPROSTIOL AND/OR ITS EQUIVALENT: 0, 1, AND 3-DAY WITHDRAWALS, PPB (MEAN±S.D.)

Sample Matrix	Withdrawal Period[a] 0-Day	1-Day	3-Day
Liver	12.78±4.17	3.66±0.41	0.92±0.44
Kidney	31.06±4.27	5.48±1.96	0.85±0.51
Fat	2.32±0.46	ND	ND
Flank Muscle	1.79[b]	ND	ND
Inj. Site #1	4.45±1.99	2.67±2.81	1.20[b]
Flank Muscle	3104.3±744.84	249.78±147.21	10.68±9.07

[a]Animals sacrificed at 12, 24, and 72 hours after the second dosage were designated as 0-day, 1-day, and 3-day, respectively.

[b]A single value. All the other values are the averages of 3-4 values.

ND-not detected.

TABLE IV. COMPARISON OF HPLC AND THERMOSPRAY/HPLC/MS RETENTION TIMES OF THE URINE EXTRACT

Peak #	HPLC Relative Retention Time (RRT)[a]	TSP/HPLC/MS RRT[b]	$[M + Ac]^-$ m/z	Molecular Weight	Spectrum #
1	0.64	-	-	-	-
2	0.67	0.70	463	404	1
3	0.72	0.71	461	402	2
4	0.78	0.79	463	404	3
5	0.80	0.81	487/533	428/474	4/5
6	0.84	0.83	463	404	6
7	0.91	0.92	495	436	7
8	1.00	1.00	517	458	8

[a]Retention time of the HPLC peak relative to HPLC peak #8.

[b]Retention time of the peak in the total ion current relative to total ion current peak #8.

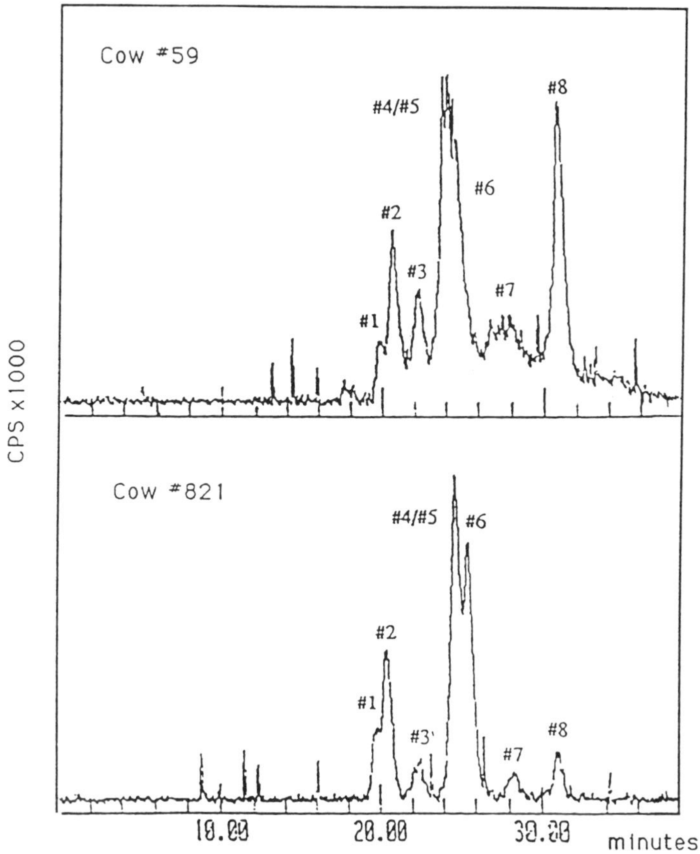

HPLC System: Waters 590 pumps, 680 controller, in/us Raytest Ramona radioactivity detector with 600 ul CaF2 solid cell. Column, Waters uBondapack C-18 (10u) 35 cm x 4.5 mm
Mobile phase system : pump A 65/35 CH3CN/0.1 M NH4Ac pump B 0.1 M NH4Ac, Gradient program (Waters 6) 15% A to 100% A over 60 minutes.

Figure 1. Comparison of Metabolic Profiles of Urine: Cow #59 and #821

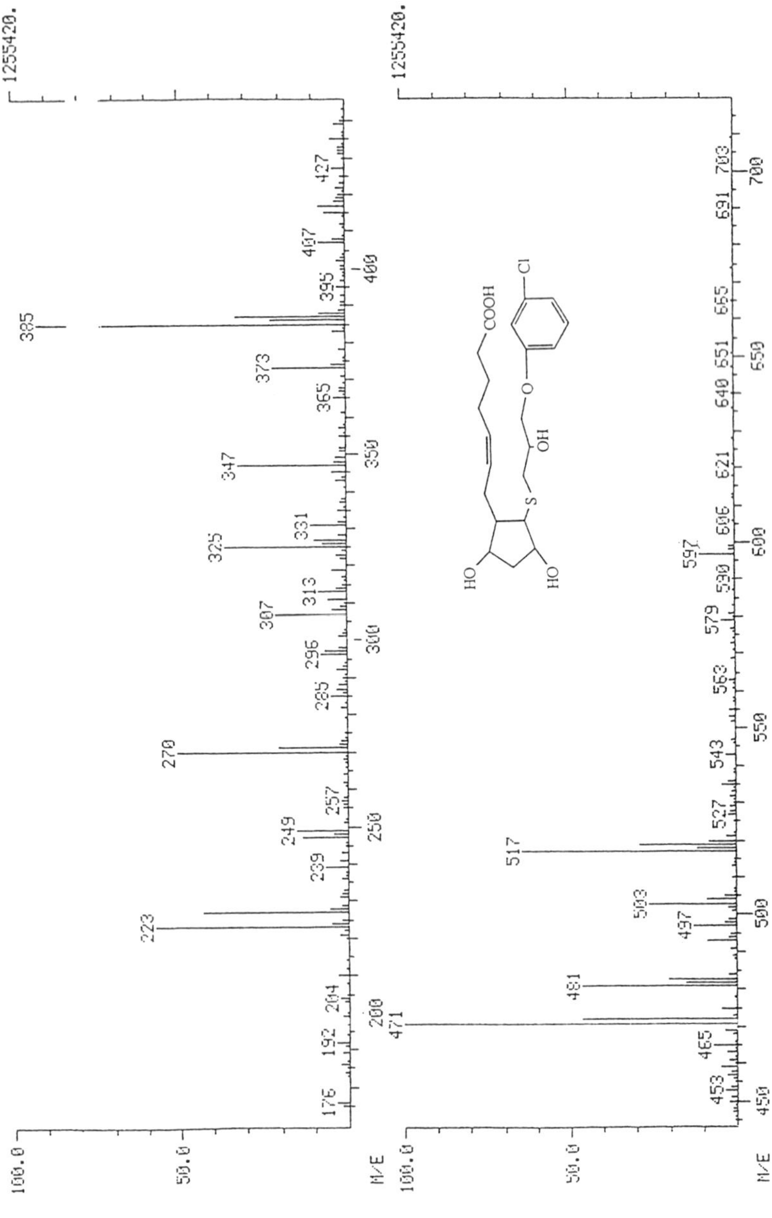

Figure 2. TSP/HPLC/MS Spectrum of Methyl Ester of Luprostiol

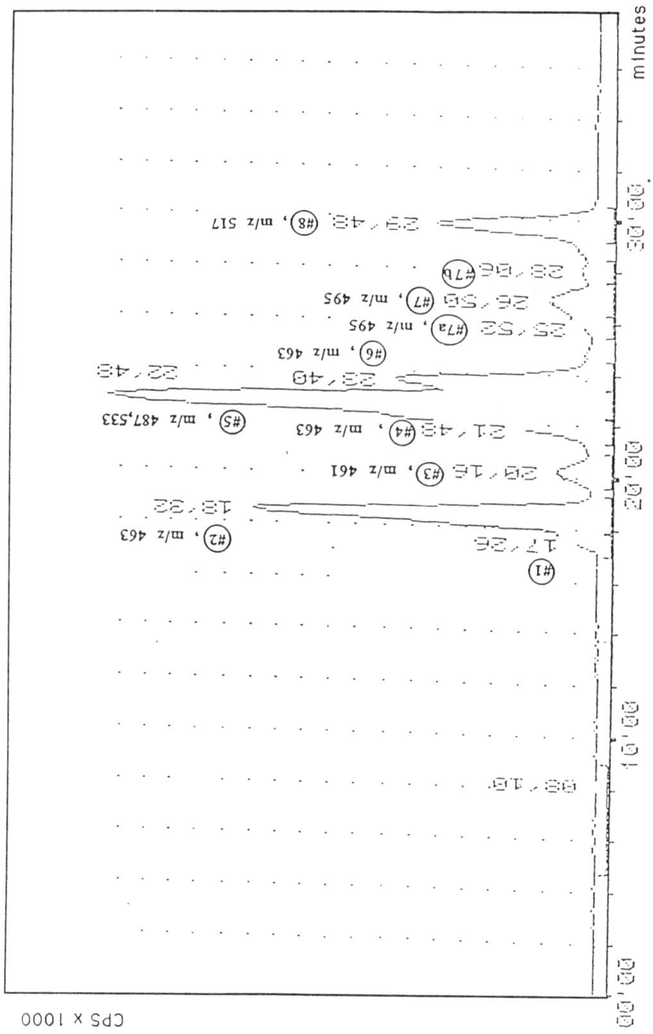

HPLC System: Waters 590 pumps, 680 controller, in/us Raytest Ramona radioactivity detector with 600 ul CaF2 solid cell.
Column: Waters uBondapack C-18 (10 u) 35 cm x 4.5 mm.
Mobile Phase System: pump A 65/35 CH3CN/0.1 M NH4Ac, pump B 0.1 M NH4Ac, gradient program (Waters 6) of 15% of A to 100% A over 60 minutes

Figure 3. Metabolic Profile of Urine (0–12 hour) After Dose #2

Peak #2 (RRT 0.67) Structure #1: The ions of m/z 463 $(M+CH_3COO)^-$, m/z 389 $(M-CH_3)^-$, m/z 373 $(M-OCH_3)^-$, m/z 345 $(M-CH_3COO)^-$, m/z 313 $(M-CH_2COOCH_3-H_2O$ $^-$ and a weak ion of m/z 445 probably due to loss of H_2O were observed. There were doublets, present at m/z 389/391, 373/375, 345/347, and 313/315 due to ^{35}chlorine and ^{37}chlorine isotopes. Based on the evidence, the structure of methyl ester of tetranor-Luprostiol (Figure 4) is proposed. The peak represented 9.54 - 17.48% of all the metabolies.

Peak #3 (RRT 0.72) Structure #2: The ions of m/z 461 $(M+CH_3COO)^-$ and m/z 257 were observed in the spectrum. Chlorine doublets were present at m/z 461/463 and 257/259. The acetate adduct (m/z 461) had two mass units less than that of the methyl ester of tetranor-Luprostiol (m/z 463). This is probably due to the oxidation of the secondary alcohol at C-9 position to form a ketone as previously observed in chlorprostinol which has a similar structure to Luprostiol except for the replacement of the S group with a C. The structure of a methyl ester of 9-ketotetranor-Luprostiol (Figure 4) is proposed for this spectrum. The peak represented 5.76 - 7.95% of all the metabolites.

Peak #4 (RRT 0.78) Structure #3: The ions of m/z 463 $(M+CH_3COO)^-$, m/z 403 $(M-H)^-$ and m/z 387 $(M-H-O)^-$ were seen with an ion corresponding to ^{37}Cl associated with all the fragments. A weak chlorine doublet was also seen at m/z 157/159 corresponding to the fragment $[CH_2-O-C_6H_3(OH)-Cl]^-$ indicating a possible hydroxylation of the aromatic ring. The fragmentation corresponded to the proposed structure of an acetate adduct of the methyl ester of δ-lactone of Luprostiol with possible oxidation of the S group and the aromatic ring (Figure 4). The peak generally co-eluted with peak #5 and both together represented 25.78 - 40.23% of all the metabolites.

Peak #5 (RRT 0.80) Structures #4 and #5: The ions m/z 487 and 533 were seen in the spectrum probably arising from the acetate adducts of two molecular ions. Both ions had chlorine doublets at m/z 487/489 and 533/535. The possible structures are given in Figure 4. Repeated attempts to separate the two adducts were not successful.

Peak #6 (RRT = 0.84) Spectrum #6: The ions of m/z 463 $(M+CH_3COO)^-$, m/z 477 $(M+CH_3COO-O)^-$, m/z 431 $(M-CH_3COO-2xO)^-$, m/z 415 $(M+CH_3COO-SO)^-$, m/z 387 $(M-H-O)^-$, m/z 371 $(M-H-2xO)^-$ and a loss of the fragment $CH_2-O-C_6H_6-Cl$ from M resulting in an ion of (m/z 263) were present in the spectrum. There were doublets at m/z 463/465, 447/449, 431/433, 415/417, 387/389 and 371/373 for chlorine isotopes. Based on the evidence, the structure of a δ-lactone of Luprostiol is proposed (Figure 4). The peak represents 17.46 - 24.27% of all the metabolites.

Peak #7 (RRT 0.91) Structure #7: The ion of m/z 495 $(M+CH_3COO)^-$ and its ^{37}Cl ion at m/z 497 were the only ions seen in the spectrum. A tentative structure of the methyl ester of the

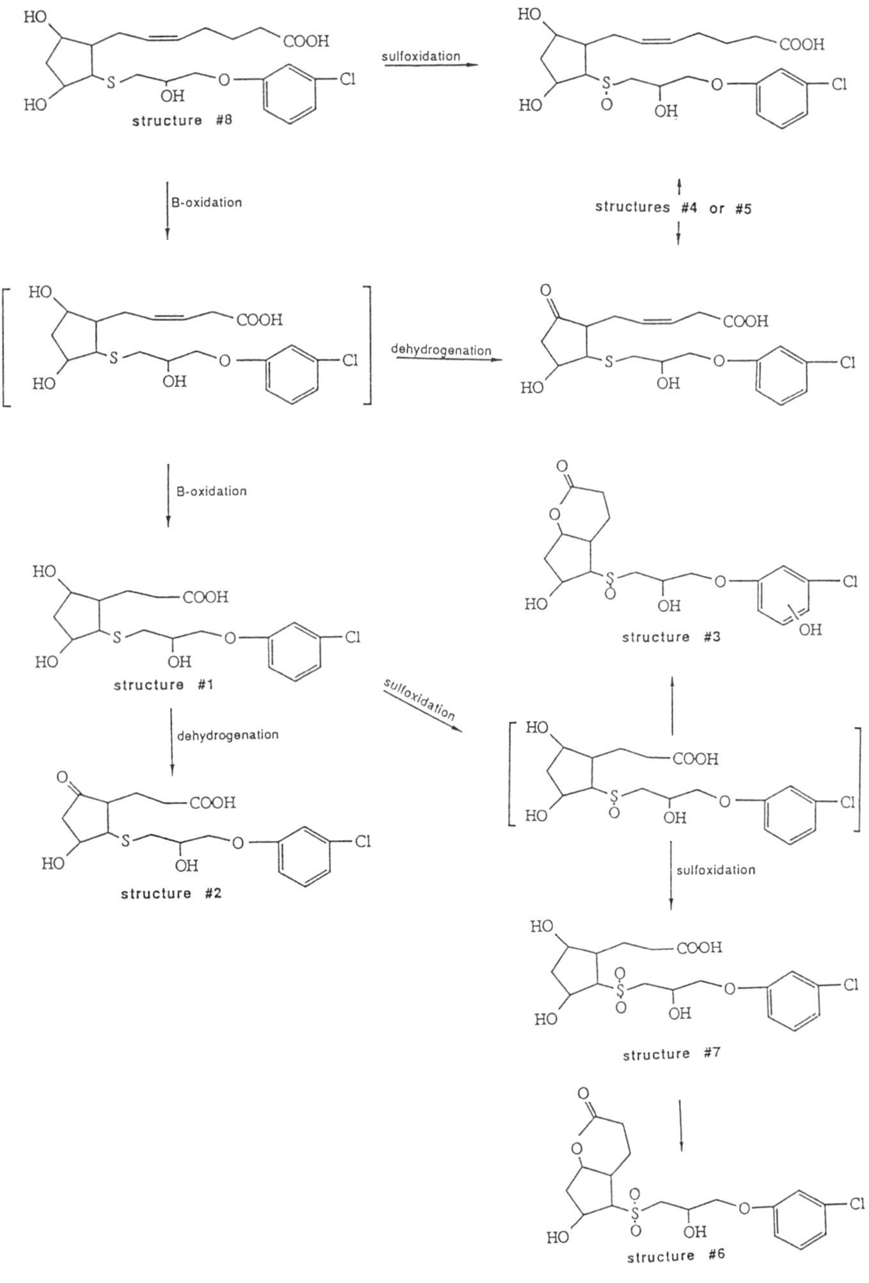

Figure 4. Possible Metabolic Pathways of Luprostiol

sulfone of tetranor-Luprostiol is proposed (Figure 4). The peak represented 2.41 - 13.60% of all the metabolites.

Peak #8 (RRT = 1.00) Structure #8: the ion of m/z 517 $(M+CH_3COO)^-$ with its characteristic chlorine doublet at m/z 517/519 indicated the structure of an acetate adduct of the Luprostiol methyl ester. This was confirmed by the analysis of a standard methyl ester of Luprostiol under similar conditions. The peak represented 4.25 - 22.57% of all the metabolites.

Comparison of Metabolic Profiles in Urine, Liver and Kidney. The metabolic profiles of liver and kidney were compared with that of urine. The comparison was done using the analysis of the extracts on two reverse phase HPLC systems. Based on the comparison of RRTs, it was evident that the metabolites found in liver and kidney were also found in the urine.

Metabolic Pathways. The major route of metabolism of Luprostiol was found to be the β-oxidation of the carboxy side chain. This has also been observed in other prostaglandin derivatives of the $PGF_{2\alpha}$ type. Carboxylic acids of dinor and tetranor structures were produced as a result of β-oxidation. Biotransformation of the cyclopentane ring at the C-9 was also found to take place to form 9-ketotetranor acid. This has been observed earlier in a similar prostaglandin derivative, clorprostenol (10). The oxidation of the secondary alcohol at C-15 was not evident and may probably be due to the presence of an oxyaryl substituent at C-16 position. Oxidation of the S group and a possible aromatic ring oxidation were observed. The metabolic pathways are illustrated in Figure 4.

Conclusions

The majority of the ^{35}S radioactivity (78 to 96%) was excreted in urine and feces within the first 48 hours after dose administration. Residues of ^{35}S-Luprostiol and/or its equivalent in milk during the first 12 hours were 1.81 ± 0.61 ppb (n=10) after a dosage of 15 mg. Less than 1 ppb of residues were detected in liver and kidney tissues of animals sacrificed 3 days after dose administration. Following a 30 mg dose, a concentration of 4.22 ± 0.95 ppb (n=4) was detected in milk during the first 12-hour sampling period. At this dosage, the residue levels in kidney and liver tissues of 0-day animals were 31.06 and 12.78 ppb, respectively. However, the residues depleted to 50% of 0-day levels in liver, kidney and injection site after 24 hours.

Luprostiol was found to be biotransformed in several metabolic routes within 12 hours after administration of a 30 mg dose. The major pathway was found to be β-oxidation of the carboxylic side chain. Metabolites in liver and kidney compared closely with those of urine based on the HPLC retention times.

Literature Cited

1. Goding, J.R.; Cumming, I.A.; Chamley, W.A.; Brown, J.M.; Cain, M.D.; Cerini, J.C.; Cerini, M.E.; Findlay, J.K.; O'Shea, J.D. and Pemberton, D.H., Gynaecol. Invest. **1971**,2, 73-97.

2. McCracken, J., Ann. N.Y. Acad. Sci. **1971** 180, 456-472.

3. Pharris, B.B. and Wyngarden, L.J., Proc. Soc. Exp. Biol. Med. **1969**,130.92-94.

4. Lager, John L., Veterinary Medicine/Small Animal Clinician, **1977**, 87-92.

5. Bourne, G.R.; Moss, S.R.; Phillips, P.J. and Shuker, B. Xenobiotica, **1979**, 9, 10, 623-631.

6. Tomlinson, R.V.;;;pires H.R. and Kent, J.S. Journal of Animal Science, **1984**, 59, 164-169.

7. Tomlinson, R.V., Spires H.R. and Bowen, T.L. J. Dairy Sci. **1985**, 68, 2072-2077.

8. Hoult, J.R.S. and Moore, P.K. BR. J. Pharmac. **1977**, 61, 615-626.

9. Bourne, G.R.; Moss, S.R.; Phillips, P.J.; and Shuker, B. Biomed. Mass Spectrom. **1980**, 7, 226-230.

10. Kaykaty, M.; Weiss, G. and Barbalas, M. J. Agric. Food Chem. **1986**, 34, 4.688-694.

11. Thornburgh, B.A.; Shaw, S.R. and Wickrema sinha, A.J. Enr. J. Drug Metab. Pharm. **1986**, 11, 1, 61-69.

12. Voyksner, R.D. and Bush, E.D. Biomed. Environ. Mass Spec. **1987**, 14, 213-220.

RECEIVED February 28, 1992

Chapter 16

Ractopamine Hydrochloride, a Phenethanolamine Repartitioning Agent

Metabolism and Tissue Residues

J. E. Dalidowicz, T. D. Thomson, and G. E. Babbitt[1]

Lilly Research Laboratories, A Division of Eli Lilly and Company, Greenfield, IN 46140–0708

Ractopamine HCl (dl-4-hydroxy-α-[[[3-(4-hydroxyphenyl)-1-methylpropyl]-amino]methyl]benzenemethanol hydrochloride) is a phenethanolamine which promotes growth and carcass leanness when fed to swine. Residue and metabolism studies were carried out using ^{14}C ractopamine HCl in swine , dogs, and rats. Swine eliminate nearly 85% of the administered ractopamine HCl during the first day resulting in relatively low tissue residues. High performance liquid chromatography of liver and kidney extracts from ^{14}C ractopamine HCl treated swine, dogs, and rats showed four major metabolites of ractopamine HCl. The metabolites were characterized as three monoglucuronides and a diglucuronide of ractopamine.

Ractopamine HCl, a phenethanolamine repartitioning agent, increases the amount of lean meat and dressing percentage and decreases the amount of carcass fat when fed to swine during the last 50 kg of gain (*1-4*). At the same time ractopamine HCl also increases the rate of weight gain and feed conversion (*5*). On the biochemical level ractopamine HCl increases nitrogen retention (*6*) protein synthesis (*7*), enhances lipolysis and suppresses lipogenesis (*8*). Recently ractopamine HCl has been shown to decrease the sensitivity and responsiveness of isolated adipocytes to insulin (*9*). In the present report, the nature of the residues in tissues and excreta of swine fed ractopamine HCl and the comparative metabolism of ractopamine HCl in swine, dogs, and rats will be described.

Materials and Methods

^{14}C Ractopamine HCl. The ^{14}C ractopamine HCl was synthesized at Lilly Research Laboratories. The material was uniformly labeled either in one or the other benzene rings. Radiochemically equivalent amounts of each label were mixed with unlabeled ractopamine HCl and used in the studies. The specific activity of the ^{14}C ractopamine HCl dose was 0.25 μCi/mg in the swine residue studies, 0.5 μCi/mg in the balance-excretion study, and 2.0 μCi/mg in the swine, dog, and rat metabolism studies.

[1]Current address: DowElanco, Greenfield Laboratories, Greenfield, IN 46140–0708

0097–6156/92/0503–0234$06.00/0

Dose and Treatment. Female (gilts) and castrated male (barrows) crossbred swine weighing approximately 45 kg were used in the studies. They were housed in individual metabolism crates and fed a ration containing 20 or 30 ppm of ^{14}C ractopamine HCl for 4-10 days. A female and a male adult Beagle dog, weighing 7-8 kg, were dosed morning, noon, and evening (6 hours apart) each day by gavage at 0.5 mg/kg of ^{14}C ractopamine HCl for four days. Female and male Fisher strain 344 rats, weighing 120-200 g, were dosed once daily by gavage at 2 mg/kg for seven days. Daily collections of feces and urine were taken as needed. At slaughter, the appropriate tissues were collected, ground wet, and analyzed.

Analysis of Samples. Tissues, feces, urine, and sample fractions were assayed for radioactivity by liquid scintillation counting as previously described (*10*). Ractopamine HCl concentration in tissues was determined by high performance liquid chromatography (HPLC) on a reverse phase phenyl column with electrochemical detection using 0.5 M $NH_4H_2PO_4$ buffer pH 4.5/CH_3CN (4:1) as the mobile phase. A 50-g ground tissue sample was homogenized twice with 100 mL of methanol. The homogenates were centrifuged and the two supernatants combined. A 10 mL aliquot of the extract was diluted with 2 mL of water, the methanol evaporated, and an additional 5 mL of water added. The aqueous sample was adjusted to pH 10.5 with 100 µL of 2 M sodium bicarbonate and extracted twice with 10 mL of ethyl acetate. The ethyl acetate was evaporated to dryness, the residue dissolved in 5 mL of acetonitrile/methanol (9:1), and applied to a Bond Elut acid washed silica cartridge (Analytichem International, Harbor City, CA). After washing the cartridge with small amounts of acetonitrile/methanol (9:1), methanol, and dichloromethane, the ractopamine HCl was eluted with 8 mL of dichloromethane/methanol/triethylamine (84:15:1). The eluate was evaporated to dryness in a rotary vacuum evaporator, the residue dissolved in 2 mL of the HPLC mobile phase, and analyzed by HPLC.

Extracts containing ractopamine metabolites were analyzed by reverse phase C_{18} HPLC with 0.05 M CH_3COONH_4 buffer, pH 4.5/CH_3CN (9:1) as the eluant. The column effluent was collected on a fraction collector. The location of the ^{14}C metabolites was determined by counting each fraction. Glucuronidase (Type IX) and sulfatase (Type VI) were obtained from Sigma Chemical Co., St. Louis, MO and used as per the company's procedures.

Residue Studies

Excretion of ^{14}C Residues. A balance-excretion study was conducted with three crossbred swine (two barrows and one gilt). The swine were fed a ration containing 20 mg/kg of unlabeled ractopamine HCl for five days. At the end of this period, each animal received a one-time dose of 40 mg of ^{14}C ractopamine HCl incorporated in the feed. The entire urinary and fecal output of each animal was collected at 24-hour intervals for seven days. The data showed that 96% of the administered dose was excreted. Of this 88% was in urine and 8% in feces. Most of the dose (95%) was excreted in the first three days with 85% eliminated during the first day. Urine was also the major route of elimination of ractopamine HCl in dogs and rats (80 and 70%, respectively).

Tissue Residues. A number of residue studies were conducted to determine the amount of and to characterize the nature of the residues in swine fed ractopamine HCl. In a steady-state study, a ration containing 30 mg/kg of ^{14}C ractopamine HCl was fed to groups of two gilts and one barrow for four and ten days and to one gilt and two barrows for seven days. The animals were sacrificed at a practical zero-time withdrawal (12 hours) after the feeding period. Muscle, liver, kidney, and backfat were collected and analyzed for total ^{14}C residues. Table I shows that the kidney and

Table I. Radioactive Residues (Mean ppm) in Tissues of Swine Dosed with 30 ppm of ^{14}C-Ractopamine HCl in the Feed

	Days Fed		
	4	7	10
Liver	0.36 ± 0.21	0.25 ± 0.04	0.42 ± 0.12
Kidney	0.52 ± 0.15	0.47 ± 0.07	0.66 ± 0.11
Muscle	0.02 ± 0.00	0.02 ± 0.00	0.02 ± 0.00
Fat	NDR[a]	NDR	NDR

[a] No detectable residue at a limit of detection of 0.015 ppm.

liver had the highest level of residues. Muscle contained very low residues while backfat had no detectable residues. Evaluation of the data using linear regression analysis showed no statistical difference between the four-, seven-, and ten-day results. The ^{14}C residue concentrations, therefore, reached steady-state in all tissues after a four day feeding of ractopamine HCl.

Three different residue depletion studies were also conducted. In the first two studies, the swine were fed a ration containing 30 mg/kg of ^{14}C ractopamine HCl for four days. In the first study, groups of one gilt and two barrows each were sacrificed after a practical zero- and four-day withdrawal periods while groups of two gilts and one barrow were sacrificed after a two- and seven-day withdrawal periods. The four pertinent tissues were collected and the ^{14}C residue concentration determined. In the second study, groups of one gilt and one barrow were sacrificed at a practical zero-, one-, and two-day withdrawal times. Besides the usual tissues, blood and bile were also collected. Table II shows a summary of the data from both studies. At the practical zero-time withdrawal period, kidney contained the highest level of ^{14}C residues. At subsequent withdrawal times, however, the liver had the highest concentration of ^{14}C residues. Muscle and fat had either extremely low or no detectable residues.

In the third residue depletion study, eight gilts and eight barrows received ^{14}C ractopamine HCl at 20 mg/kg in the feed for seven days. Groups of three gilts and three barrows were sacrificed after one- and two-day withdrawal periods and the remaining two gilts and two barrows after a three-day withdrawal period. After each sacrifice the ^{14}C residue concentration in muscle, liver, kidneys, and backfat and the ractopamine HCl concentration in liver and kidneys was determined. The mean ^{14}C residue and ractopamine HCl concentrations in liver and kidney are shown in Table III. Muscle and backfat had no detectable ^{14}C residues at all the withdrawal times. The depletion of the ^{14}C residues in liver proceeded quite rapidly falling below 90 µg/kg (90 ppb) between the first and second day of ractopamine HCl withdrawal. The concentrations of ractopamine after two and three days of withdrawal periods were extremely low (2-8 ppb) in both liver and kidney.

Metabolism Studies

Metabolites in Excreta. Diethylether extracts of urine at pH 10 from the second residue depletion study contained, depending on the sample, various amounts of ractopamine HCl. The urine collected the first day in five of the animals had 4% or less of ractopamine HCl of the total radioactivity percent while the urine of the sixth had 18%. By day 4 of treatment, the concentration of ractopamine HCl in the urine of all animals increased to 36-85% of the total ^{14}C residue. When the same urine samples were incubated with β-glucuronidase that did not have any sulfatase activity , over 90% of the radioactive residues were extracted into diethylether, 88-90% of which was ractopamine HCl. On the other hand, when the urine samples were incubated with sulfatase only small increases in the radioactivity of diethylether extracts were seen. Most of the non-diethylether extractable metabolitics in urine were, therefore, glucuronides. Since there are three hydroxy groups in ractopamine HCl, the possibility exists of three different monoglucuronide conjugates, as well as diglucuronides, and possibly a triglucuronide.

In order to determine the identity of the ractopamine glucuronides, the urine of the animal with the highest percentage of conjugates was further analyzed. The aqueous pH 10 fraction from the organic extraction was chromatographed on an Amberlite XAD-2 column and sequentially eluted with water, water/methanol (1:1) and methanol. The water/methanol eluate was chromatographed on a low pressure silica gel column using a nonlinear gradient of methylene chloride and methylene chloride/ethanol (9:1, 4:1, and 7:3). The radioactive fraction that eluted from the silica

Table II. Depletion of Radioactive Residues (Mean ppm) in Tissues, Blood, and Bile of Swine Dosed with 30 ppm of ^{14}C-Ractopamine HCl in the Feed

	Withdrawal Time (Days)				
	0	1	2	4	7
Liver	0.38 ± 0.15	0.17 ± 0.03	0.09 ± 0.03	0.05 ± 0.03	0.06 ± 0.03
Kidney	0.54 ± 0.19	0.13 ± 0.01	0.06 ± 0.01	0.03 ± 0.01	0.02 ± 0.01
Muscle	0.02 ± 0.01	0.01 ± 0.00	NDR[a]	0	0
Fat	0.02 ± 0.01	0.01 ± 0.00	NDR	NDR	NDR
Blood	0.04 ± 0.00	0.01 ± 0.00	NDR	NA[b]	NA
Bile	1.56 ± 0.09	0.35 ± 0.10	0.07 ± 0.07	NA	NA

[a] No detectable residues at a limit of detection of 0.01 ppm.
[b] Not analyzed.

gel column with methylene chloride/ethanol (7:3) was subjected to reverse phase C_{18} HPLC. The elution profile of the HPLC (Figure 1) shows four peaks of radioactivity which are labeled Metabolites A, B, C, and D, respectively. Subsequently, reverse phase HPLC was used to isolate Metabolites A, B, and C in sufficient quantities for spectral analysis. Mass spectroscopy employing fast atom bombardment (FAB) in glycerol containing small amounts of oxalate showed that the three metabolites had MH^+ ions of 478 m/z in agreement with the MH^+ of ractopamine glucuronide.

Proton NMR spectra were used to confirm the structure of the metabolites and establish the position of the glucuronide in Metabolites A, B, and C. The proton NMR spectra of the three metabolites as compared to ractopamine showed the appearance of complex multiplet proton resonances between 3 and 4 ppm which also occur in the spectrum of glucuronic acid confirming the conjugate nature of the metabolites. The NMR spectra of the three metabolites also showed that the α-hydroxy group is not involved in these conjugations since the proton resonance of that group was present in the spectrum of the three metabolites. Using homonuclear decoupling experiments and comparison of NMR shift data in DMSO-d_6 with ractopamine and deshydroxy ractopamine, the position of the glucuronide in each metabolite was identified. The NMR spectrum of ractopamine showed four pairs of "doublet" resonances at 6.70, 6.78, 7.00, and 7.18 ppm. Irradiation of each doublet established that the 6.78/7.18 pairs of doublets came from the protons of one aromatic ring while the 6.70/7.00 pairs of doublets came from the protons of the other aromatic ring. The NMR spectrum of deshydroxy ractopamine showed that the 6.78/7.18 doublets were shifted upfield by 0.11 and 0.15 ppm, respectively. Since the protons in Ring A (Table IV) would be affected most by the loss of the α-hydroxy group, the 6.78/7.18 ppm doublets come from the resonance of the protons in Ring A.

In the spectrum of Metabolite A, the aromatic protons at 6.70 ppm were shifted upfield by 0.22 ppm, while the resonance from the other doublets remained very close to that of ractopamine and deshydroxy ractopamine. A similar upfield shift in the resonance of the 6.70 ppm protons were obtained in the NMR spectrum of Metabolite B. Since the changes in the resonance of aromatic protons occurred in Ring B, the glucuronide is attached to Ring B in both metabolites. The difference in the two metabolites is stereoisomerism. In the NMR spectrum of Metabolite C, an upfield shift of 0.20 ppm of one of the proton doublets in Ring A can be seen. The glucuronide, therefore, is attached to Ring A. Unlike the diastereomers of Ring B glucuronides, the ones on Ring A were not separated by HPLC. Metabolite D could not be isolated in suitable quantity from swine urine to be identified. In a related project, however, a metabolite with the same retention time as Metabolite D on reverse phase HPLC was isolated and purified from bile of cattle dosed with ^{14}C ractopamine HCl. FAB mass spectrum of the isolate showed an MH^+ ion of 653 corresponding to the diglucuronide of ractopamine.

Tissue Metabolites. Ground liver or kidney tissue (25-50 g) was extracted once with 0.1 M NH_4HCO_3, pH 10 and twice with 0.1 M NH_4HCO_3, pH 10/methanol (1:1). The supernatants from the last two extractions were combined and the methanol evaporated on a rotary vacuum evaporator. The resultant aqueous solution was combined with the first extract and chromatographed on Amberlite XAD-2 as described for urine. The water-methanol and methanol fractions were combined and the methanol rotoevaporated. The resulting aqueous solution was adjusted to pH 10 and extracted with diethylether to remove ractopamine. The aqueous phase was evaporated on a rotary vacuum evaporator to dryness and analyzed by reverse phase C_{18} HPLC. Figure 2 shows that both the liver and kidneys contain the same four major metabolites (A, B, C, and D) as swine urine.

Table III. Depletion of Radioactive (RA) Residues and Ractopamine (in ppb) in Tissues of Swine Dosed with 20 ppm of ^{14}C-Ractopamine HCl in the Feed

Withdrawal (Days)	Liver RA Residues	Liver Ractopamine	Kidney RA Residues	Kidney Ractopamine
1	106 ± 30	14.8 ± 6.7	116 ± 14	32.1 ± 15.1
2	73 ± 28	3.7 ± 1.9	48 ± 7	8.3 ± 2.5
3	56 ± 10	1.7 ± 0.9	36 ± 1	3.4 ± 1.8

Table IV. Aromatic Proton NMR Resonances of Ractopamine HCl, Its Metabolites, and Deshydroxyractopamine

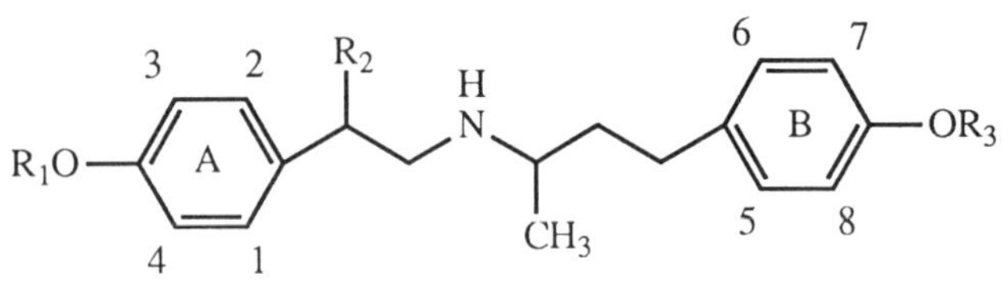

	δ (ppm) of Aromatic protons: Ring A H-1,2	Ring A H-3,4	Ring B H-5,6	Ring B H-7,8
Ractopamine	7.18	6.78	7.00	6.70
Deshydroxy-ractopamine	7.03	6.67	7.00	6.70
Metabolite A	7.08	6.71	7.04	6.92
Metabolite B	7.10	6.69	6.98	6.89
Metabolite C	7.21	6.98	6.95	6.65

Ractopamine HCl (R_1,R_3 = H, R_2 = OH). The other compounds have the following differences: deshydroxyractopamine (R_2 = H), Metabolite A and Metabolite B (R_3 = glucuronide), and Metabolite C (R_1 = glucuronide).

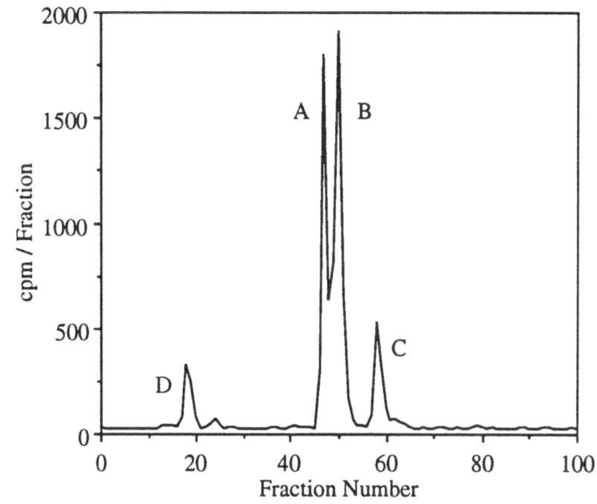

Figure 1. Reverse Phase HPLC of Urine Extract From Swine Fed ^{14}C-Ractopamine HCl. The Identity of Metabolites A,B,C, and D is Shown in Table IV.

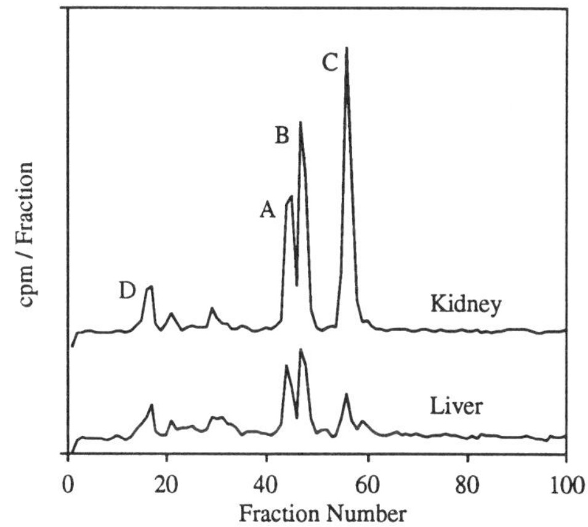

Figure 2. Reverse Phase HPLC of Liver and Kidney Extracts From Swine Fed ^{14}C-Ractopamine HCl.

Table V. Amount of Ractopamine HCl and Its Metabolites in Liver and Kidneys of Swine, Dogs, and Rats

	Percent of Total Residues					
	Liver			Kidney		
	Pig	Dog	Rat	Pig	Dog	Rat
Ractopamine	28.7	8.4	31.6	23.4	20.7	18.9
Metabolite A	7.9	6.4	12.0	11.0	7.4	29.8
Metabolite B	10.4	10.7	10.6	13.2	11.4	32.2
Metabolite C	4.6	23.9	7.0	22.3	25.4	4.9
Metabolite D	5.0	9.8	11.8	6.1	6.0	5.6
Unidentified [a]	8.1	11.3	7.6	4.2	22.7	3.4
Nonextractable	8.5	13.4	5.6	3.2	2.0	2.5

[a] Comprises 4-6 minor metabolites. Amberlite XAD-2 water eluate, which was not characterized, and losses during purification accounted for the rest of the total residues.

Comparative Metabolism. Extracts of dog and rat urine, liver, and kidneys, obtained by the same procedures as the swine extracts, were analyzed by reverse phase C_{18} HPLC. The data showed that urine and both tissues of dogs and rats contained the same four major metabolites A, B, C, and D as swine urine and tissues. Table V contains comparative quantitative data on ractopamine HCl and its metabolites in liver and kidneys of swine, dogs, and rats. The amounts as a percent of total residues in all fractions of ractopamine HCl and the four metabolites were almost the same in the three species in both liver and kidney tissue. The notable differences were that dog liver contained less ractopamine and more Metabolite C, and rat kidney contained more Metabolites A and B and less Metabolite C than the other species. The nonextractable ^{14}C residues were low in the two tissues of all species.

Conclusions. Swine eliminated ractopamine very rapidly resulting in relatively low tissue residues. After the withdrawal of ractopamine HCl from the diet, the residues depleted rapidly from all tissues examined and were no greater than 90 ppb in liver by two days after withdrawal. Muscle and fat did not have any detectable residues. The concentration of ractopamine in liver and kidneys was extremely low. Urine, liver, and kidneys each contained the same four major metabolites of ractopamine: three monoglucuronides and a diglucuronide. The same four metabolites in approximately the same percentages were found in the liver and kidneys of dogs and rats. The animals chosen for toxicological studies, therefore, were exposed to the same metabolites as those found in the edible tissues of swine.

Acknowledgments. The authors wish to thank J. L. Occolowitz for MS analyses; N. H. Terando for the synthesis of ^{14}C ractopamine HCl; T. D. Macy for the analysis of ractopamine HCl in swine liver and kidneys; and G. W. Bewley, D. R. Breese, and J. M. Marsh for their technical assistance.

Literature Cited

(*1*) Anderson, D.B.;Veenhuizen, E.L.; Waitt, W.P.; Paxton, R.E.; Young, S.S. *Fed. Proc.* **1987**, *46*, p. 1021.
(*2*) Hancock, J.D.; Peo, E.R.; Lewis, A.J.; Parrott, J.C. *J. Anim. Sci.* **1987**, *65*, (Suppl. 1), p. 309.
(*3*) Prince, T.J.; Huffman, D.L.; Brown, P.M.; Gillespie, J.R. *J. Anim. Sci.* **1987**, *65*, (Suppl. 1), p. 309.
(*4*) Anderson, D.B.; Veenhuizen, E.L.; Waitt, W.P.; Paxton, R.E.; Mowrey, D.H. *J. Anim. Sci.* **1988**, *66*, (Suppl. 1), p. 279.
(*5*) Veenhuizen, E.L.; Schmiegel, K.K.; Waitt, W.P.; Anderson, D.B. *J. Anim. Sci.* **1987**, *65*, (Suppl. 1), p. 130.
(*6*) Anderson, D.B.; Veenhuizen, E.L.; Waitt, W.P.; Paxton, R.E.; Mowrey, D.H. *J. Anim. Sci.* **1987**, *65*, (Suppl. 1), p. 130.
(*7*) Bergen, W.G.; Johnson, S.E.; Skjaerlund, D.M.; Babiker, A.S.; Ames, N.K.; Merkel, R.A.; Anderson, D.B. *J. Anim. Sci.* **1989**, *67*, pp. 2255-2262.
(*8*) Merkel, R.A.; Dickerson, P.S.; Johnson, S.E.; Burkett, R.L.; Burnett, A.L.; Bergen, W.G.; Anderson, D.B. *Fed. Proc.* **1987**, *46*, p. 1177.
(*9*) Hausman, D.B.; Martin, R.J,; Veenhuizen, E.L.; Anderson, D.B. *J. Anim. Sci.* **1989**, *67*, pp. 1455-1464.
(*10*) Magnussen, J.D., Dalidowicz, J.E.; Thomson, T.D.; Donoho, A.L. *J. Ag. Food Chem.* **1991**, *39*, pp. 306-310.

RECEIVED March 20, 1992

Indexes

Author Index

Affiliation Index

Subject Index

Production: Margaret J. Brown
Indexing: Deborah H. Steiner
Acquisition: Anne Wilson and A. Maureen Rouhi
Cover design: Sarah Sonyong Chung

Printed and bound by Maple Press, York, PA

Highlights from ACS Books

Good Laboratory Practices: An Agrochemical Perspective
Edited by Willa Y. Garner and Maureen S. Barge
ACS Symposium Series No. 369; 168 pp; clothbound, ISBN 0–8412–1480–8

Silent Spring Revisited
Edited by Gino J. Marco, Robert M. Hollingworth, and William Durham
214 pp; clothbound, ISBN 0–8412–0980–4; paperback, ISBN 0–8412–0981–2

Insecticides of Plant Origin
Edited by J. T. Arnason, B. J. R. Philogène, and Peter Morand
ACS Symposium Series No. 387; 214 pp; clothbound, ISBN 0–8412–1569–3

Chemistry and Crime: From Sherlock Holmes to Today's Courtroom
Edited by Samuel M. Gerber
135 pp; clothbound, ISBN 0–8412–0784–4; paperback, ISBN 0–8412–0785–2

Handbook of Chemical Property Estimation Methods
By Warren J. Lyman, William F. Reehl, and David H. Rosenblatt
960 pp; clothbound, ISBN 0–8412–1761–0

The Beilstein Online Database: Implementation, Content, and Retrieval
Edited by Stephen R. Heller
ACS Symposium Series No. 436; 168 pp; clothbound, ISBN 0–8412–1862–5

Materials for Nonlinear Optics: Chemical Perspectives
Edited by Seth R. Marder, John E. Sohn, and Galen D. Stucky
ACS Symposium Series No. 455; 750 pp; clothbound; ISBN 0–8412–1939–7

Polymer Characterization:
Physical Property, Spectroscopic, and Chromatographic Methods
Edited by Clara D. Craver and Theodore Provder
Advances in Chemistry No. 227; 512 pp; clothbound, ISBN 0–8412–1651–7

From Caveman to Chemist: Circumstances and Achievements
By Hugh W. Salzberg
300 pp; clothbound, ISBN 0–8412–1786–6; paperback, ISBN 0–8412–1787–4

The Green Flame: Surviving Government Secrecy
By Andrew Dequasie
300 pp; clothbound, ISBN 0–8412–1857–9

For further information and a free catalog of ACS books, contact:
American Chemical Society
Distribution Office, Department 225
1155 16th Street, NW, Washington, DC 20036
Telephone 800–227–5558